Satellite Altimetry - Theory, Applications and Recent Advances

Edited by Tomislav Bašić

Published in London, United Kingdom

Satellite Altimetry - Theory, Applications and Recent Advances
http://dx.doi.org/10.5772/intechopen.102197
Edited by Tomislav Bašić

Contributors
Eman A. Alshari, Bharti W. Gawali, Tarek M. El-Geziry, Sajjad Sajjadi, Zdenek Martinec, Georgios S. Vergos, Eleni A. Tzanou, Dimitrios A. Natsiopoulos, Tomislav Bašić, Ljerka Vrdoljak

First published in London, United Kingdom, 2023 by IntechOpen
IntechOpen is the global imprint of INTECHOPEN LIMITED, registered in England and Wales, registration number: 11086078, 5 Princes Gate Court, London, SW7 2QJ, United Kingdom

British Library Cataloguing-in-Publication Data
A catalogue record for this book is available from the British Library

Additional hard and PDF copies can be obtained from orders@intechopen.com

Satellite Altimetry - Theory, Applications and Recent Advances
Edited by Tomislav Bašić
p. cm.
Print ISBN 978-1-80355-879-0
Online ISBN 978-1-80355-880-6
eBook (PDF) ISBN 978-1-80355-881-3

Meet the editor

Professor Tomislav Bašić obtained an MSc in Surveying Engineering from the University of Zagreb, Croatia, in 1980 and a Ph.D. in Geodetic Science from Leibniz University Hannover, Germany, in 1989. During his postdoctoral research at the Department of Geodetic Science and Surveying, Ohio State University, Columbus, USA, Dr. Bašić worked on the TOPEX/POSEIDON three-year prime mission research project for NASA. His research interests include physical geodesy, the Earth's gravity field, satellite altimetry, geodetic reference frames, and the application of global navigation satellite systems (GNSS) technology for the establishment of fundamental geodetic networks and geo-kinematic studies.

Contents

Preface

Satellite radar altimetry, with its global coverage and level of precision, is a key technique for observing the ocean surface and many aspects of land surfaces, over the last three decades. Through interdisciplinary collaboration, satellite altimetry has integrated the efforts of international researchers to demonstrate the technique itself, previous and future missions, and measurements and their accuracy, including orbit determination and ocean circulation models. Satellite altimetry has revolutionized our understanding of the Earth's sea-level shape and its change over time, enabling the monitoring of natural and human-induced water mass balance change, marine gravity computations, high-resolution seafloor bathymetry reconstruction, a better understanding of plate tectonics, and more. For researchers in geodesy and geophysics, oceanography, and other space and earth sciences, satellite radar altimetry is critical for unifying vertical height systems, regional and global geoid, gravity, and bathymetry modelling, and monitoring sea-level rise, vertical land motion, and ice sheet melting.

This book provides a state-of-the-art overview of satellite radar altimetry technology and recent developments in its use. It discusses applied studies utilizing altimeter data to determine sea-level changes and sea-level variability, marine bathymetry modelling from satellite altimeter-derived gravity data and the related issue of solving terrain corrections for gravity reductions, as well as the possibility of using artificial intelligence (AI) in monitoring Earth's changes, including the use of satellite altimetry data.

The introductory chapter, "Satellite Altimetry – Overview", by the editor, provides a geodetic review of the basic principles and formulas of satellite radar altimetry technology and corrections for the processing and application of satellite altimetry data, primarily in geodetic tasks. Along with the evolution of technology and guidelines for future development, the chapter presents selected products that open a wide area for future scientific and applied research in the field of geosciences.

Chapter 2, "Cyclo-Stationarity in Sea Level Variability from Satellite Altimetry Data and Correlation with Climate Indices in the Mediterranean Sea" by Dimitrios A. Natsiopoulos, Eleni A. Tzanou, and Georgios S. Vergos, deals with numerical experiments for the statistical analysis of Sea Level Anomaly (SLA) variations in the Mediterranean obtained from the exact repeat missions of Envisat, Jason-1, Jason-2, and Cryosat-2 data for the period between 2002 and 2016. Analysis of the empirical covariance functions of SLAs show that there is a significant annual variation that is evident for the entire period under study. This variation is in line with the thermal expansion of the sea due to increasing temperatures during the summer and early fall months and lower temperatures during winter. Moreover, the seasonal cycle can also be attributed to atmospheric forcing due to the variation in atmospheric pressure in the Mediterranean. To identify possible correlations with global and regional climatic phenomena that influence the ocean state, three indexes have been investigated: the

Southern Oscillation Index (SOI), the Mediterranean Oscillation Index (MOI), and the North Atlantic Oscillation (NAO).

Chapter 3, "Sea-Level Changes", by Tarek M. El-Geziry, examines the problem of the rate of sea level rise (SLR) on a global and relative level using complementary mareograph tide gauge measurements and satellite altimetry measurements. Numerous investigations have shown that the rise of the mean sea level was about 1.3 mm/year during the 20th century. During 2006–2018, the sea level on a global scale has risen as much as 3.7 mm/year, but it also changed on a relative (regional/local) level. There are two main reasons for this: (1) the thermal expansion of seawater due to climate change and global warming, and (2) the melting of ice in the Arctic and Antarctic regions. Both are direct consequences of climate change, which is mainly caused by anthropogenic activities. Additionally, vertical ground movement (subsidence/uplift) can affect the calculated relative SLR rates. Given that nature has changed significantly over the last several decades, the necessity of international cooperation, public awareness campaigns, better monitoring tools, numerical models for simulations and predictions, and the further development of satellite technology such as satellite altimetry is specifically highlighted.

Chapter 4, "Bathymetry Estimation from Satellite Altimeter-Derived Gravity Data", by Ljerka Vrdoljak and Tomislav Bašić, gives an overview of the physical problem of and different approaches to estimating bathymetry from satellite altimeter-derived gravity data. It also discusses the possibility of regional bathymetry modelling in the Adriatic Sea by the gravity-geologic method (GGM) in the space domain with a simpler algorithm, higher resolution, and satisfactory quality as compared to global solutions. Comparison with modern shipborne bathymetric surveys shows that bathymetry estimated from altimetry has a coarse spatial resolution and lesser accuracy, especially in coastal areas. As compared to chart soundings, all models had the least accuracy in the coastal area shallower than 20 m. The quality increased up to 10% of the depth in the deepest parts of the Adriatic. Limitations of the bathymetry estimated from altimetry can be overcome by using more available high-quality bathymetry in important coastal areas.

Chapter 5, "Terrain Corrections in Gravity and Gradiometry" by Sajjad Sajjadi, Zdenek Martinec, and Patrick Prendergast, deals with determining the influence of topography in the calculation of physical parameters, especially the surface of the geoid, which must be defined before the calculation of the short-wave and long-wave components of the calculated physical parameters both from terrestrial data, where satellite altimeter-derived gravity data can be counted to, and from the recent satellite missions, that is, gravity measurements (GRACE mission) and gradiometry (GOCE mission). Since the corrections enter the first of three steps of the Remove–Compute–Restore (RCR) procedure for applying the Stokes integral, this study focuses on determining these corrections. The effects formulation is introduced, and the effects are calculated on high-elevation topography in Ireland using Helmert's second condensation method. Finally, the chapter examines the effects of topography on determining the heights of the geoid.

Chapter 6, "Artificial Intelligence Techniques for Observation of Earth's Changes", by Eman A. Alshari and Bharti W. Gawali, highlights that the art and science of measuring the planet through sensors or satellites, known as remote sensing, together with

GIS technology, is essential for collecting data about the Earth. The chapter presents an overview of the application of AI in detecting Land Use Land Cover (LULC) changes on Earth. It also includes a short subsection on the use of satellite altimetry to monitor changes in mean sea level. The chapter provides an overview of Earth observation satellites and their development, reviews AI procedures (supervised and unsupervised methods) for LULC, highlights the fundamentals of Machine Learning (ML) classifiers, and presents the challenges of AI techniques for LULC classification.

Finally, I sincerely thank all the chapter authors for their excellent contributions.

Tomislav Bašić
University of Zagreb,
Zagreb, Croatia

Chapter 1

Introductory Chapter: Satellite Altimetry – Overview

Tomislav Bašić

1. Introduction

Radar satellite altimetry is one of the basic satellite measurement techniques intended primarily for solving global geodetic tasks by means of radar measurements from satellites toward the Earth. Satellite altimetry ensures the collection of high-precision global data of uniform accuracy on sea level, which enables monitoring of the geophysical characteristics of the sea and larger water surfaces, that is, marine topography and circulation within liquid water bodies. During the last four decades, satellite altimetry has revolutionized geosciences, especially oceanography, geophysics, and geodesy. This measurement method found its application in modeling the shape of the Earth and the Earth's field of acceleration of gravity, modeling the relief of the seabed and vertical displacements of the Earth's crust in coastal areas, and monitoring climate phenomena and long-term climate changes. Satellite altimetry data is distributed in the form of original measurements and products ready for use in geosciences, most often calculated models, or calculation services. This chapter presents the fundamental principles of the radar altimetric measurement method, its historical development, achievements, and expected improvements in technology soon, as well as the scientific and professional results achieved so far in the development and application of technology.

2. Evolution of technology

The concept of satellite altimetry was developed in the sixties of the twentieth century as part of NASA's (National Aeronautics and Space Administration) NGSP (National Geodetic Satellite Program) initiative for the development of a space geodetic program and was formalized in 1969 during a conference on solid Earth and ocean physics [1]. During that time of conceptual development, satellite altimetry's main goal was to determine the Earth's shape, which today can really be considered a limited ambition [2].

Following the timeline of technology development, the operational characteristics of the satellite missions launched to date are given in **Table 1**. The variety of satellite missions is shown regarding the height of the satellite orbits, the inclination angles of the orbits, the coverage of the geographical area of the Earth, the distances of the satellite paths on the Equator, and the frequency band and altimeter frequencies.

IntechOpen

Mission	Orbit height (km)	Inclination	Latitude coverage	Equator track distance (km)	Band	Frequency (GHz)
GEOSAT	785	108°	72°	163	Ku	13.5
ERS-1/2	785	98°	81°	80	Ku	13.8
TOPEX/ POSEIDON Jason-1/2/3 Sentinel-6	1336	66°	66°	315	Ku/C	13.6/5.3
GFO	785	108°	72°	163	Ku	13.5
Envisat	785	98°	81°	163	Ku/S	13.6/3.2
CryoSat-2	717	92°	88°	7	Ku	13.6
HY-2A/2B	964	99°	60°	90	Ku/C	13.6/5.3
SARAL/ ALTIKA	800	98°	81°	90	Ka	35

Table 1.
Operational characteristics of satellite altimetry missions.

2.1 Development phase of technology

With the launch of the Skylab satellite in 1973, the development and experimental era of satellite altimetry began, revealing the great potential of radar altimetry. Already in 1975, the GEOS-3 (Geodynamics Experimental Ocean Satellite) satellite was launched, as the third satellite in the NGSP program [3]. The satellite mission achieved measurement precision in a one-second interval of about 25 cm, which, along with the low accuracy of determining the satellite's orbit of about 5 m, could not meet the requirements for application in geodetic purposes [4].

Significant progress was made during the SEASAT satellite mission, which achieved 5-centimeter precision of altimetry measurements, and methods of determining satellite orbits and geophysical corrections applied during altimetry data processing were significantly improved [5]. This was achieved by simultaneous observation with different instruments from the satellite: (1) a SAR instrument (Synthetic Aperture Radar), (2) a scattering meter, which was used to determine the wind speed and direction above the water surface, (3) a multi-frequency microwave radiometer, which was used for determining water surface temperature, and (4) radar altimeter [4]. The data thus collected enabled the modeling of the circulation of ocean waves and wind along the sea surface, the geoid model in the sea area, and the topography of the seabed.

The last mission of the development phase, GEOSAT (GEOdetic SATellite), was launched in 1985 with the basic goal of determining the potential of the acceleration of gravity at sea and modeling the topography of the sea primarily for the needs of the US Navy [6]. The success of the satellite mission is evidenced by the calculation that the satellite mission saved the Navy in the amount of about 280 million US dollars by replacing the long-term shipboard gravimetric measurement [7].

Practical calculations of gravity anomalies and geoid undulations from satellite altimetry data began already during the development phase. As an example, we cite the usage of GEOS-3, SEASAT, and GEOSAT altimetry data and ETOPO5U bathymetry data to globally determine gravity anomalies and sea surface heights in the ocean areas, using the least squares collocation method [8, 9].

2.2 The modern and future era of technology

The 1990s represent a turning point in the application of technology, during which significant advances were made toward operational oceanography, that is, the possibility of forecasting sea level and temperature and sea currents, for which the basic prerequisites are reliable and high-frequency measurements with altimeters that are available in real-time. The potential of the technology was also recognized by the European Space Agency (ESA), which defined the basic goals of future missions in the PRARE (Precision Range and Range-Rate Equipment) project: (1) calibration of radar altimeters within 10 cm using laser retroreflectors on Earth, (2) download and distribution of measured data in real-time, and (3) automation of data processing and development of rapidly available standardized products [10], which began to be realized with the ERS-1 (European Remote Sensing) satellite mission. **Figure 1** displays a view of the current and future altimetry constellations [11].

At the same time, NASA, and the French agency CNES (Centre National d'Etudes Spatiales) developed and launched TOPEX/Poseidon, one of the most significant satellite geodetic missions in history that revolutionized satellite altimetry as a technology [12, 13]. The radar altimeter of the TOPEX/Poseidon satellite, in addition to the standard microwave frequency of about 12 GHz (Ku band), was also equipped with another frequency in the C-band (about 5 GHz), which became the standard for later satellite missions. This, along with the introduction of the third frequency on the satellite's microwave radiometer, enabled the efficient calculation of the ionospheric correction and the removal of the influence of wind on altimetry measurements [2]. This satellite mission achieved the accuracy of the determination of the satellite's orbit expressed by the root mean square error of about 2.5 cm and the measurement precision with a standard deviation of 2 cm, which achieved the estimated precision of

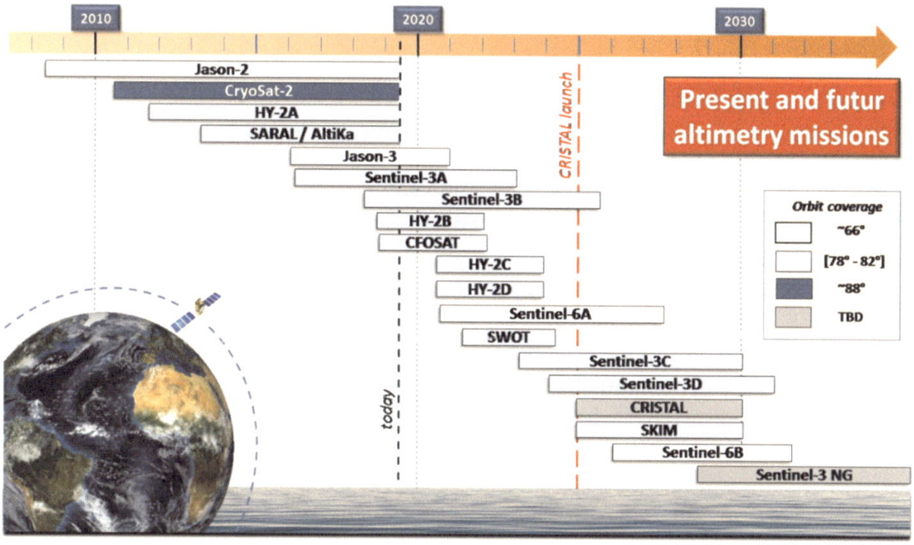

Figure 1.
Present and future altimetry missions [11].

determining the sea level in the open sea by satellite altimetry of about 4 cm [14]. Considerable progress in the accuracy of satellite orbit determination was achieved with the DORIS measuring system for tracking satellites from Earth.

During the nineties, the ERS-2 mission (the successor to ERS-1) and the GFO (Geosat Follow-On), which succeeded the GEOSAT mission, were launched. A similar intensity of the frequency of launching altimetry satellite missions continued in the first years of the twenty-first century, when the JASON-1 (Joint Altimetry Satellite Oceanography Network) mission succeeded TOPEX/Poseidon, with a four-year period of simultaneous observation of the two satellites. JASON-1 was designed primarily to determine trends in mean sea level change, or to assess the consequences of climate change [15]. In 2002, the Envisat satellite mission was launched, which succeeded the ERS missions, with the basic goal of enabling the creation of environmental studies, the development of biological oceanography studies, and the mapping of ice surfaces on Earth [16].

The following satellite missions are currently active [17]: CryoSat-2 (Cryogenic Satellite), a mission created as part of ESA's space program in 2010, HY-2a (Haiyang), mission realized in 2011 under the leadership of CNSA (China National Space Administration), SARAL, a mission launched in 2013 with the cooperation of ISRO (Indian Space Research Organization) and CNES, SENTINEL-3, a mission launched in 2015 as part of ESA's Copernicus space program with the leadership of EUMETSAT, JASON-3, a designed in collaboration of NASA and ESA as the successor of TOPEX/Poseidon and Jason 1/2, HY-2b (Haiyang), launched as the second in the series of Chinese Haiyang satellites in 2018, and Sentinel-6 (previously referred to as Jason CS), launched in late 2020, which continues the EU Copernicus and NASA program and previous TOPEX/Poseidon and Jason 1/2/3 satellite missions.

Satellite altimetry missions of the near future should enable and improve the efficient monitoring of the surface level of lakes, rivers, and coastal areas, forecast the intensity of tropical cyclone disturbances, hurricanes, and enable the development of improved models of sea changes and currents. In the future, the launch of SWOT (Surface Water Ocean Topography) satellite mission is expected, primarily to enable terrestrial water monitoring.

3. Principle of satellite altimetry measurement

Satellite altimetry is a method of determining the height of the sea surface in relation to a defined geodetic reference frame (ellipsoid or geoid) based on the measurement of the distance between the satellite and the instantaneous sea surface. Distance measurement is based on the measurement of the time required for the radar signal to travel from the satellite to the water surface and from the water surface to the satellite, with a series of corrections necessary due to signal propagation through the atmosphere and the influence of geophysical phenomena of water surfaces on the reflected signal [2]. A prerequisite for determining the height of the water surface is knowledge of the position of the satellite in a clearly defined geodetic reference system.

3.1 Basic formulas and corrections of satellite altimetry data

The altimeter from the satellite sends a short pulse of microwave radiation of known power toward the sea surface at regular time intervals. The pulse interacts with the rough sea surface and part of the radiation is reflected toward the altimeter. With the known speed of the radar signal, that is, the speed of light in a vacuum c, and neglecting refraction, the distance of the satellite from the sea surface Robs can be calculated based on the time t required for the signal pulse to travel twice [14]. The fundamental equation of altimetry then takes the form:

$$R_{obs} = \frac{ct}{2} \tag{1}$$

After applying the corrections, the basic equation for calculating the corrected distance R has the form:

$$R = R_{obs} - \sum_i \Delta R_i = R_{obs} - \left(\Delta R_{tdry} + \Delta R_{twet} + \Delta R_{iono} + \Delta R_{dyn} + \ldots\right), \tag{2}$$

where the values ΔR_i, $i = 1\ldots$ represent corrections, ΔR_{tdry} and ΔR_{twet} are the influence of the dry and wet components of the troposphere, ΔR_{iono} is the influence of the ionosphere, physical influences on the surface of the sea such as sea currents and tides, motion of the Earth's pole, solid Earth tides and the dynamics of the sea (sea state bias) ΔR_{dyn}.

The consequence of the latter can most often be expressed by the slope of the waves that cause the radar signals to bounce with a small displacement, which is related to local conditions (wind and waves). An offset correction due to sea position is attempted to account for the difference between the scattered surface and the true mean sea level within the altimeter footprint. The correction is a combined effect of electromagnetic and asymmetric shifts [2].

The most common corrections and models used today in the processing of original altimetric measurements, namely orbit, dry/wet tropospheric correction, ionospheric correction, inverse barometric correction, Sea tides and solid Earth tides, and geodetic reference surface can be found in example [18]. All influences on the signal cause a delay in the return signal, so the corrections are positive amounts (Eq. (2)). The distance estimate is variable along the satellite orbit due to changes in sea surface topography and changes in the height of the orbit relative to the center of the Earth.

The basic prerequisite for determining the height of the water surface based on the measured distance from the satellite is the knowledge of the position of the satellite flying in a predefined orbit, that is, its height h_{OR} in a fixed geocentric reference system. Determining the position of satellites today is done through DORIS and SLR tracking from Earth and GNSS technology from space. The estimated accuracy of determining the orbit of modern satellites is 1 cm. The sea surface height (SSH), or the water surface in general, in the reference system in which the position of the satellite is expressed (as a rule, the ellipsoidal height) can be expressed as:

$$SSH = h_{OR} - R = h_{OR} - R_{obs} + \sum_i \Delta R_i. \tag{3}$$

The dynamic effects of geostrophic sea currents are of primary importance for satellite altimetry applications in oceanography [14]. Therefore, the dynamic topography of the sea (SST; **Figure 2**) can be represented mathematically by the equation:

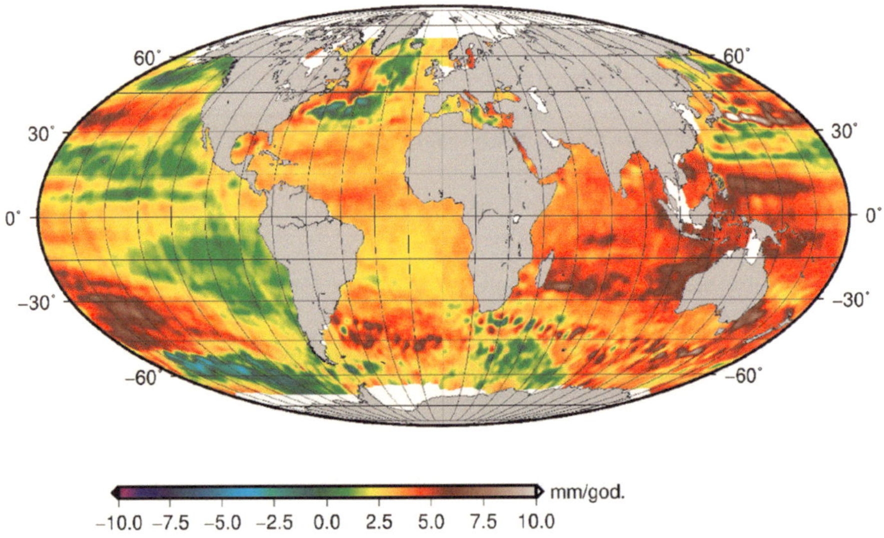

Figure 2.
Global trend of mean sea level change based on CU data [19], adjusted by the method of least squares for the period 1993–2012.

$$SST = SSH - N - h_T - h_A, \tag{4}$$

where N is the undulation of the geoid, and h_T and h_A are the influence of sea tides or atmospheric pressure, which are calculated from the model or corrections for the moment of measurement. The influence of atmospheric pressure can be described by the inverse barometric (IB) effect, that is, the direct influence of changes in atmospheric pressure on the level of the water surface, whereby the level of the water surface decreases with an increase in air pressure and vice versa. Changes in air pressure of 100 Pa cause changes of about 1 cm in the water surface level. The average monthly influence of atmospheric pressure on sea level in the Mediterranean area is about 3 cm [20].

By determining the average change in SSH heights over a period, we can also define the mean sea surface MSS. The level of the mean sea level above the geoid is called the mean dynamic ocean topography which provides data on the magnitude of ocean circulations. To define the MSS size, it is important to use the latest data due to the development of radar altimeters themselves, precise calculation of orbits, geophysical corrections, and slow changes in ocean currents. The mean sea level is a surface created by averaging a longer period of sea level observations lasting a minimum of one year due to pronounced seasonal effects of sea level change, and optimally 18.6 years due to the Earth's nutation period [21].

The sea level can also be expressed by the sea level anomaly SLA in relation to the mean sea surface MSS:

$$SLA = SSH - MSS. \tag{5}$$

The mean level shown on the geoid represents the topography of the sea (sea surface topography, [22]), which is one of the basic variables in determining the model of the shape of the Earth—the geoid.

3.2 Advanced altimeter processing methods

Satellite altimetry provides centimeter accuracy over the open ocean from orbits more than 1000 km above Earth. However, the estimation of the height of the water level is significantly less accurate in coastal and inland areas, mainly due to difficulties in estimating tropospheric corrections, high-frequency atmospheric signals, tides, and mostly problems related to land contamination in radar altimeter footprints [23].

Improved altimetry data in coastal areas and on land is obtained through retracking —signal reprocessing procedures using complex algorithms. Problems related to data processing procedures are explained in detail in Ref. [2]. Today, the most used data processing algorithms in the coastal area are ALES (Adaptive Leading Edge Subwaveform) and X-TRACK [24, 25]. Altimetry retrackers compare the wave strength of return signals most often with previously known wave models, and in this way, reconstruct measurements of water surfaces [24]). ALES+ was later designed for sea ice, coastal and inland waters [26], while Goddard Space Flight Center (GSFC) designed several retrackers for ice regions. Thanks to such retrackers, altimeter errors are reduced, ensuring coverage and use of satellite altimetry in coastal zones and inland water areas. All retracked data is available through the coastal altimetry community.

The most significant recent development in satellite altimetry technology has been the introduction of Delay-Doppler (DD) or SAR-mode altimetry, which allows better observation of small-scale features (below 50 km) and improved spatial resolution along the satellite track compared to conventional pulse-limited altimeters [27]. DD altimetry uses the Doppler effect, caused by satellite motion in the along-track direction, to improve spatial resolution in the same direction and thus enable along-track data sampling (e.g., up to 300 m for Sentinel-3). In other words, the footprint of the DD altimeter is reduced by an order of magnitude compared to conventional altimeters (from several kilometers to several hundred meters) [28]. Therefore, DD altimetry, such as those on CryoSat-2 (SIRAL, SAR Interferometric Radar Altimeter) and Sentinel-3 (SRAL, Synthetic Radar Altimeter), provide more and/or improved data over the ocean, especially in sea ice areas and coastal areas.

4. Altimeter products

4.1 Sea-level change

The use of satellite altimetry data to monitor changes in the mean sea level, as a basic geodetic task, is one of the most effective ways of monitoring climate change. The assessment of the change in mean sea level today indicates a trend of increase of 3.1 ± 0.4 mm/year (**Figure 2**), which is agreed upon by five leading scientific institutions in the field of research: AVISO, CSIRO (Commonwealth Scientific and Industrial Research Organisation), CU (University of Colorado Boulder), GSFC (Goddard Space Flight Center), and NOAA, whose trend estimates are in good agreement, although the data processing and trend calculation methods differ.

4.2 Gravity model

With satellite altimetry, it is possible to calculate the mean sea level in relation to the geoid, which enables efficient and high-quality modeling of the geoid surface, especially for the sea and ocean area [8, 29].

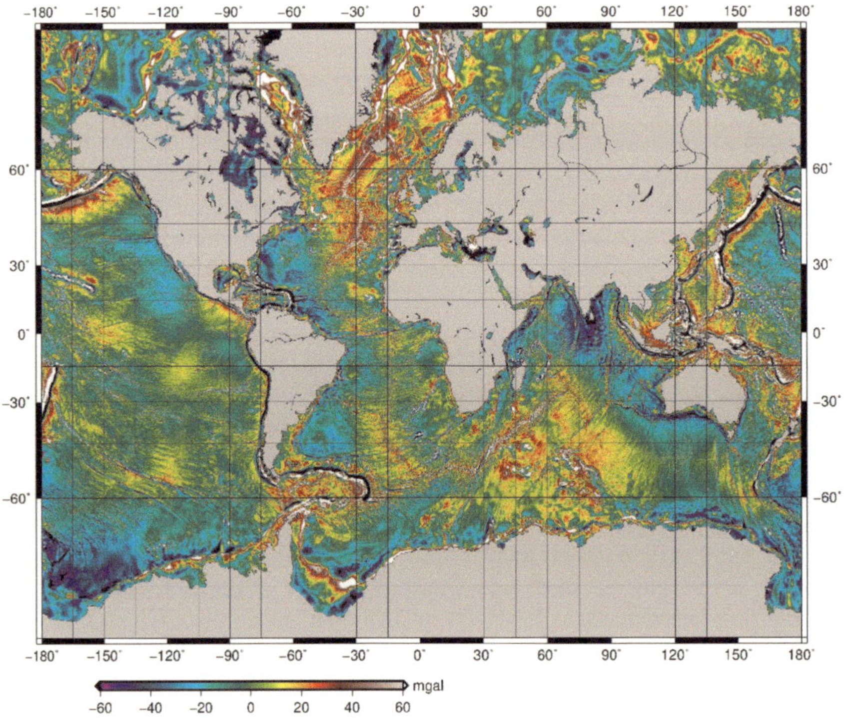

Figure 3.
Altimetry-derived global ocean gravity map DTU15 [30].

From the satellite altimetry missions data, it is possible to reconstruct the gravity field on water surfaces. Usually, three procedures for calculating the anomalies of the Earth's gravity acceleration field from altimetric data are used: (1) by applying the least squares collocation method with altimetric data and calculated surface slopes along the satellite path, (2) by applying the least squares collocation method with altimetry data and calculated vertical deflections, and (3) by applying the Vening-Meinesz formula to the vertical deflection data on water surfaces obtained from altimetry. **Figure 3** shows the global model of free air anomalies DTU15, calculated from several altimetry missions [30].

4.3 Bathymetry

Satellite altimetry data regularly distributed over seas and oceans can be combined with infrequent and relatively expensive depth measurements by ultrasonic depth sounders at specific ship locations to produce bathymetric models [31]. Maps created in this way cannot be used for precise underwater navigation, but they can indicate the topography of the seabed, that is, larger geo-tectonic structures, such as lithospheric plate boundaries, etc. **Figure 4** shows a depth map derived from altimetry data [32].

Calculated gravity acceleration anomalies and bathymetric maps are very often used in interdisciplinary research related to geodesy—for example, tectonophysics, and studies on tectonic movements because they effectively reveal the boundaries of tectonic plates and the specificities of local areas. In addition, satellite altimetry data are successfully used for applications in oceanography and glaciology [14], but also in multidisciplinary early warning systems, such as those predicting floods or tsunamis

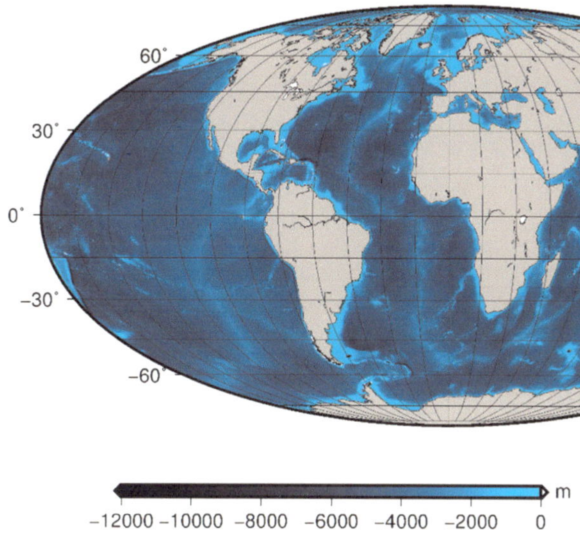

Figure 4.
Global depth model created based on altimetry data [32].

and other climate-related forecasting systems, which lead to operational oceanography, i.e. a system of predicting sea-related variables, such as sea level, temperature, and currents, based on long-term routine measurements and real-time observations of the ocean and atmosphere.

In Ref. [33], the conjoint analysis of vertical land motion of the Dubrovnik area was derived from the ESA's Sentinel-1 InSAR data, continuous GNSS observations, and differences in the sea-level change obtained from all available satellite altimeter missions for the Dubrovnik area and tide gauge measurements.

5. Conclusion

Since the early nineties of the twentieth century, satellite altimetry has been applied in various geodetic and interdisciplinary research. Satellite altimetry enables advanced determination of the Earth's shape by implementing a model of the acceleration of the Earth's gravity. Furthermore, satellite altimetry enables efficient, global, and relatively frequent monitoring of changes in mean sea level as an indicator of climate change and serves as a basis for the establishment of height systems at sea and on land. Finally, as part of interdisciplinary research, satellite altimetry enables the determination of global depth models, the assessment of the impact of sea level changes, vertical movements of the Earth's crust in coastal areas, and obtaining the tectonic geomorphology of the seabed.

Author details

Tomislav Bašić[1,2]

1 University of Zagreb, Croatia

2 Faculty of Civil Engineering, Architecture and Geodesy, University of Mostar, Bosnia and Hercegovina

*Address all correspondence to: tomislav.basic.smb@gmail.com

References

[1] Kaula WM. The Terrestrial Environment. [Washington]. Springfield, Va: National Aeronautics and Space Administration; for Sale by the Clearinghouse for Federal Scientific and Technical Information; 1970

[2] Vignudelli S, Kostianoy AG, Cipollini P, Benveniste J. Coastal Altimetry. Berlin: Springer Science Business Media; 2011

[3] Stanley H. The GEOS 3 project. Journal of Geophysical Research: Solid Earth. 1979;**84**(B8):3779-3783

[4] Rosmorduc V, Benveniste J, Lauret O, Maheu C, Milagro M, Picot N. Radar Altimetry Tutorial. Bruxelles: ESA; 2011. pp. 112-128

[5] Tapley BD, Born GH, Parke ME. The Seasat altimeter data and its accuracy assessment. Journal of Geophysical Research: Oceans. 1982;**87**(C5):3179-3188

[6] McConathy DR, Kilgus CC. The Navy GEOSAT mission: An overview. Johns Hopkins APL Technical Digest. 1987; **8**(2):170-175

[7] Sandwell DT, Müller RD, Smith WH, Garcia E, Francis R. New global marine gravity model from CryoSat-2 and Jason-1 reveals buried tectonic structure. Science. 2014;**346**(6205):65-67

[8] Bašić T, Rapp RH. Oceanwide Prediction of Gravity Anomalies and Sea Surfacc Heights Using Geos-3, Seasat and Geosat Altimeter Data and ETOPO5U Bathymetric Data. Columbus/ Ohio: Department of Geodetic Science and Surveying, Ohio State University, Report No. 416, 1-89; 1992

[9] Rapp RH, Bašić T. Oceanwide gravity anomalies from GEOS-3, Seasat and Geosat altimeter data. Geophysical Research Letters. 1992;**19**(19): 1979-1982

[10] Strawbridge F, Laxon S. ERS-1 altimeter fast delivery data quality flagging over land surfaces. Geophysical Research Letters. 1994;**21**(18):1995-1998. DOI: 10.1029/94GL01730

[11] Available from: https://www.polarmonitoringproject.org/2025-2030-altimetry-constellation/

[12] Cazenave A, Dominh K, Ponchaut F, Soudarin L, Cretaux JF, Le Provost C. Sea level changes from Topex-Poseidon altimetry and tide gauges, and vertical crustal motions from DORIS. Geophysical Research Letters. 1999;**26**(14):2077-2080

[13] Nerem RS. Measuring global mean sea level variations using TOPEX/ POSEIDON altimeter data. Journal of Geophysical Research: Oceans. 1995;**100** (C12):25135-25151

[14] Fu LL, Cazenave A. Satellite Altimetry and Earth Sciences: A Handbook of Techniques and Applications. Vol. 69. Cambridge: Academic Press; 2000

[15] Ménard Y, Fu LL, Escudier P, Parisot F, Perbos J, Vincent P, et al. The Jason-1 mission special issue: Jason-1 calibration/validation. Marine Geodesy. 2003;**26**(3–4):131-146

[16] Frappart F, Calmant S, Cauhopé M, Seyler F, Cazenave A. Preliminary results of ENVISAT RA-2-derived water levels validation over the Amazon basin. Remote Sensing of Environment. 2006; **100**(2):252-264

[17] Grgić M, Bašić T. Radar Satellite Altimetry in Geodesy – Theory,

Applications and Recent Developments. Open Access Peer-reviewed Chapter. In: Erol B, Erol S, editors. Geodetic Sciences – Theory, Applications and Recent Developments. London, United Kingdom: IntechOpen; 2021. pp. 77-94. DOI: 10.5772/intechopen.87679

[18] Grgić M. The estimation of the Adriatic Sea Level variability and changes based on multi-satellite altimetry data [PhD thesis]. Zagreb: University of Zagreb, Faculty of Geodesy; 2017

[19] Masters D, Nerem RS, Choe C, Leuliette E, Beckley B, White N, et al. Comparison of global mean sea level time series from TOPEX/Poseidon, Jason-1, and Jason-2. Marine Geodesy. 2012;**35**:20-41

[20] Grgić M, Nerem RS, Bašić T. Absolute sea level surface modeling for the Mediterranean from satellite altimeter and tide gauge measurements. Marine Geodesy. 2017;**40**(4):239-258

[21] Dehant V. On the nutations of a more realistic Earth model. Geophysical Journal International. 1990;**100**(3): 477-483

[22] Koblinsky CJ, Nerem RS, Williamson RG, Klosko SM. Global scale variations in sea surface topography determined from satellite altimetry. In: Sea Level Changes: Determination and Effects. American Geophysical Union, Geophysical Monograph Series. Vol. 60. John Wiley & Sons. 1993. pp. 155-165

[23] Dufau C, Martin-Puig C, Moreno L. User requirements in the coastal ocean for satellite altimetry. In: Vignudelli S, Kostianoy A, Cipollini P, Benveniste J, editors. Coastal Altimetry. Berlin, Heidelberg: Springer; 2011:51-60. DOI: 10.1007/978-3-642-12796-0_3

[24] Passaro M, Cipollini P, Vignudelli S, Quartly G, Snaith H. ALES: A multi-mission adaptive subwaveform retracker for coastal and open ocean altimetry. Remote Sensing of Environment. 2014; **145**:173-189. DOI: 10.1016/j.rse.2014. 02.008

[25] Roblou L, Lyard F, Le Henaff M, Maraldi C. X-TRACK, a new processing tool for altimetry in coastal oceans. In: Proceedings of 2007 IEEE International Geoscience and Remote Sensing Symposium, Barcelona, Spain, 2007. pp. 5129-5133

[26] Passaro M, Rose S, Andersen O, Boergens E, Calafat F, Dettmering D, et al. ALES+: Adapting a homogenous ocean retracker for satellite altimetry to sea ice leads, coastal and inland waters. Remote Sensing of Environment. 2018;**211**: 456-471. DOI: 10.1016/j.rse.2018.02.074

[27] Santos-Ferreira A, da Silva J, Srokosz M. SAR-mode altimetry observations of internal solitary waves in the Tropical Ocean part 2: A method of detection. Remote Sensing. 2019;**11**(11): 1339. DOI: 10.3390/rs11111339

[28] Egido A, Smith W. Fully focused SAR altimetry: Theory and applications. IEEE Transactions on Geoscience and Remote Sensing. 2017;**55**(1):392-406. DOI: 10.1109/tgrs.2016.2607122

[29] Pavlis NK, Holmes SA, Kenyon SC, Factor JK. The development and evaluation of the Earth Gravitational Model 2008 (EGM2008). Journal of Geophysical Research: Solid Earth. 2012; **117**(B4):B04406

[30] Andersen O, Knudsen P, Kenyon S, Factor J, Holmes S. Global gravity field from recent satellites (DTU15) — Arctic improvements. First Break. 2017; **35**(12):37-40. DOI: 10.3997/ 1365-2397.2017022

[31] Smith WH, Sandwell DT. Bathymetric prediction from dense satellite altimetry and sparse shipboard bathymetry. Journal of Geophysical Research: Solid Earth. 1994;**99**(B11): 21803-21824

[32] Andersen O, Knudsen P, Stenseng L. The DTU13 MSS (Mean Sea Surface) and MDT (Mean Dynamic Topography) from 20 Years of Satellite Altimetry. Berlin, Heidelberg: Springer; 2015

[33] Grgić M, Bender J, Bašić T. Estimating vertical land motion from remote sensing and In-situ observations in the Dubrovnik area (Croatia): A multi-method case study. Remote Sensing. 2020;**12**(21):3543. DOI: 10.3390/rs12213543

Chapter 2

Cyclo-Stationarity in Sea Level Variability from Satellite Altimetry Data and Correlation with Climate Indices in the Mediterranean Sea

Dimitrios A. Natsiopoulos, Eleni A. Tzanou and Georgios S. Vergos

Abstract

The exploitation of altimetric datasets from past and current satellite missions is crucial to both oceanographic and geodetic applications. For oceanographic studies, they allow the determination of sea level anomalies as deviations from a static mean sea level. This chapter deals with numerical experiments for the statistical analysis of Sea Level Anomaly (SLA) variations in the Mediterranean. SLA empirical covariance functions were calculated to represent the statistical characteristics of the sea variation for the period between 2002 and 2016. The variation of monthly SLA time series was investigated, and a correlation analysis was performed in terms of epoch-based pattern re-occurrence. To identify possible correlations with global and regional climatic phenomena that influence the ocean state, three indexes have been investigated, namely the Southern Oscillation Index (SOI), the Mediterranean Oscillation Index (NOI), and the North Atlantic Oscillation (NAO). Finally, Empirical Orthogonal Functions (EOF) and Principal Component Analysis (PCA) were applied to all SLA time series and for each satellite mission to extract the individual dominant modes of the data variability. After the analysis, the SLA field is separated into spatial structures (EOF modes) and their corresponding amplitudes in time, the Principle Components (PCs).

Keywords: SLA, EOF, variance, MOI, NAO, SOI, variability, cyclo-stationarity

1. Introduction

Fluctuations in the sea level is a significant problem, as the sea level continues to rise dramatically [1, 2]. Regarding Europe, sea level rise is very important, as especially in the Mediterranean Sea, the majority of the population is concentrated along the coastline [3–6]. Almost 529 million citizens live around Mediterranean countries, with 205 million (~39%) on the northern shore and 324 million on the southern and eastern shores. According to existing demographic projections, this number can reach as high as 611 million within next three decades. By 2025, one-third

IntechOpen

of the Mediterranean population will live on the northern shore and two-thirds on the southern and eastern shores [7].

The population growth rate in coastal areas is accelerating, and increasing tourism adds to the pressure on the environment mainly by increasing the vulnerability of coastal ecosystems. Countryside and residences are changed, affected and destroyed by either direct human intervention or as a byproduct of the increasing anthropogenic activity in coastal areas. As sea level rises, increased salinity in groundwater could affect access to drinking water and agricultural inputs. Flooding and the destruction of infrastructure in food-producing areas is a likely result of sea level rise. Coastal erosion with environmental, social and economic impact will result in repercussions that will affect large coastal population. Tourism industry with all its activities (hotels, transports etc), a sector that several economies depend on can be lost forever [8–17].

The global sea level level is rising extremely fast. Several studies during the last twenty years have shown that new records of sea level are set with the projections for the future to be even more discouraging. As a result, sea level variations with time and their projections are of high importance. Global tides-gauge data over the last century showed a rise of 18.5 cm [18], while similar studies using only tide-gauge data from coastal areas have been done for the Mediterranean [15, 19–21]. On the other hand, studies using annually averaged satellite altimetry data from various missions show a somewhat higher rate (3 to 7 mm) of rising than tide gauge (TG) data [22–25].

Fluctuations in water temperature, salt content and added water volume especially from melting glaciers contribute to the short and long-term variations. The significant positive trend in sea level rise rely mainly on global Earth's temperature increase and on thermal expansion of Earth itself [26]. Therefore, observing the fluctuations of sea level along with the trends is crucial as it can be a decisive factor in coastal areas tracking system for their decisive management.

Satellite altimetry offer high accuracy and resolution information for sea surface heights for almost the last half century, increasing the in-situ sea level measurements from tide-gauge found both on offshore and coastal areas. Altimetry focused satellites, with the low resolution mode (LRM) data from the first missions to the last synthetic aperture radar (SAR) observations from Cryosat-2 and Sentinel satellites increased the surface observations, resulting in a constant global coverage of sea. The concept of this technique is based on the transmission of the altimeter sensor, located on the bottom part of the satellite, of a microwave or laser pulse to the sea surface and its reflection of water. The distance between the satellite and the sea-surface is given by the distance formula, by multiplying the speed of light by the two-way travel time (see **Figure 1**). Taking into account the satellite altitude from a surface (ellipsoid), Sea Surface Heights (SSH) are available for time of the measurement. This is the main advantage of satellite altimetry, as it provides real time and precise information over seas. Moreover, the short, repeated period of the latest altimetry missions tend to be a very useful tool for the estimation of the sea level rise as increases the temporal sampling rate of sea surface heights measurements.

Many studies have been conducted in the past concerning the calculation of mean sea surface (MSS) models from altimetry data with of these studies presenting global models [27, 28] or regional ones [29–32]. Compared to other older techniques like tide gauge or shipborne data collection, satellite altimetry can provide high-resolution and precision information both for ocean and geophysics and several studies including gravity, dynamic ocean topography, circulation, temperature etc. rely on altimeter data [33–35].

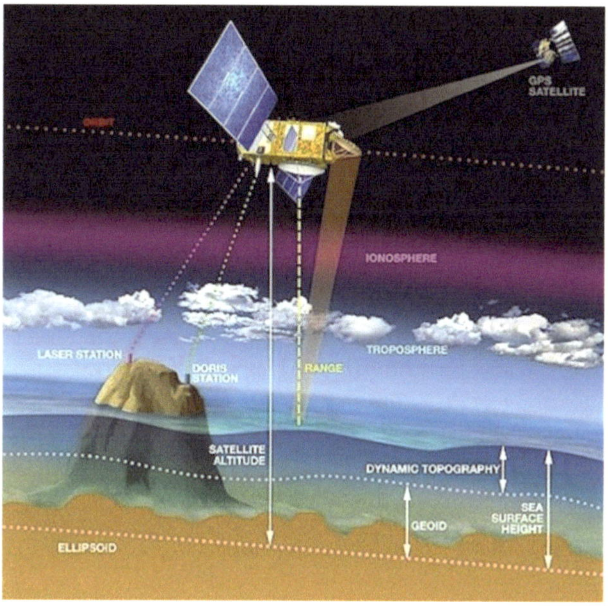

Figure 1.
Principle of satellite altimetry (credit AVISO, 2021).

Based on the previously mentioned, this study aims to analyze the available Envisat, Jason-1, Jason-2 and Cryosat-2 data in the form of Sea Level Anomalies (SLAs) spanning the Mediterranean Sea. Although Mediterranean basin is nearly enclosed by land, mainly on the north (Europe) and on the south (Africa), variations of sea level can be analyzed as the altimetry data cover a very long period of repeated measurement ground footprints. By studying the empirical covariance functions computed for all altimetry missions available for this study, spanning the years from 2002 to 2016, variations of SLAs can be further analyzed and useful conclusions on sea level variation can be defined. Based on this, results confirming the cyclo-stationarity of the SLAs can be inferred and connected with climate change indices over the oceans. To this extent, three such indexes have been investigated, namely the Southern Oscillation Index (SOI), the Mediterranean Oscillation Index (NOI), and the North Atlantic Oscillation (NAO). Finally, the last step in this study, involves the Principal Component Analysis (PCA). PCA employed the data of each satellite mission separately or the whole data as well. Through this method, dominant patterns can be extracted, identifying thus either annual or seasonal modes, signal or noise.

2. Area under study, available data and pre-processing

As already mentioned, the area under study covers the whole Mediterranean Sea, from Europe to Africa on the north–south direction and from strait of Gibraltar to Levant on the West -East direction with four altimetry missions studied in this work, Envisat, Jason-1, Jason-2, and Cryosat-2. For Jason-1, data from the various phases of the mission have been used. These comprise of data from phase A, from 15/1/2002 (cycle 1) to 26/1/2009 (cycle 259) resulting in a total number of 689,680 observations; data from phase B, during the period from 10/2/2009 (cycle 262) to 3/3/2012 (cycle 374) (512,996 observations) and from the geodetic phase (phase C), during the period from 7/5/2012 (cycle 382) to 21/6/2021 (cycle 425) and a total number of 120,973 observations (see **Figure 2** for the Jason –1 data distribution).

During phase A, each Jason-1 cycle consists of 254 passes, with almost 20% of those having available observations in the Mediterranean Sea within the satellite's period of 10 days. In phase B, the new orbit was adjacent to that of Jason-2 and covered the oceans every 4.5 days while in phase C, the satellite was set in a drifting orbit to perform a geodetic, i.e., higher resolution, mapping. For Envisat satellite, 881,657 point values (see **Figure 3**) compose phase A, within the period 14/5/2002 and 22/10/2010 (cycle 6 to cycle 77) and 150,435 points phase B with cycles 95 to 113. The values mesh is much denser than Jason-1 and is consists of 1003 passes with four times less cross-track spacing at the equator (~75 to ~300 km).

The orbit of Jason-2 was the same as that of the Jason-1 and Topex/Poseidon satellites. Data collection started with cycle 0 on 4/7/2008, and phase A ended on 2/10/2016 with cycle 303. This 8 year period of data collection resulted in 1,061,379

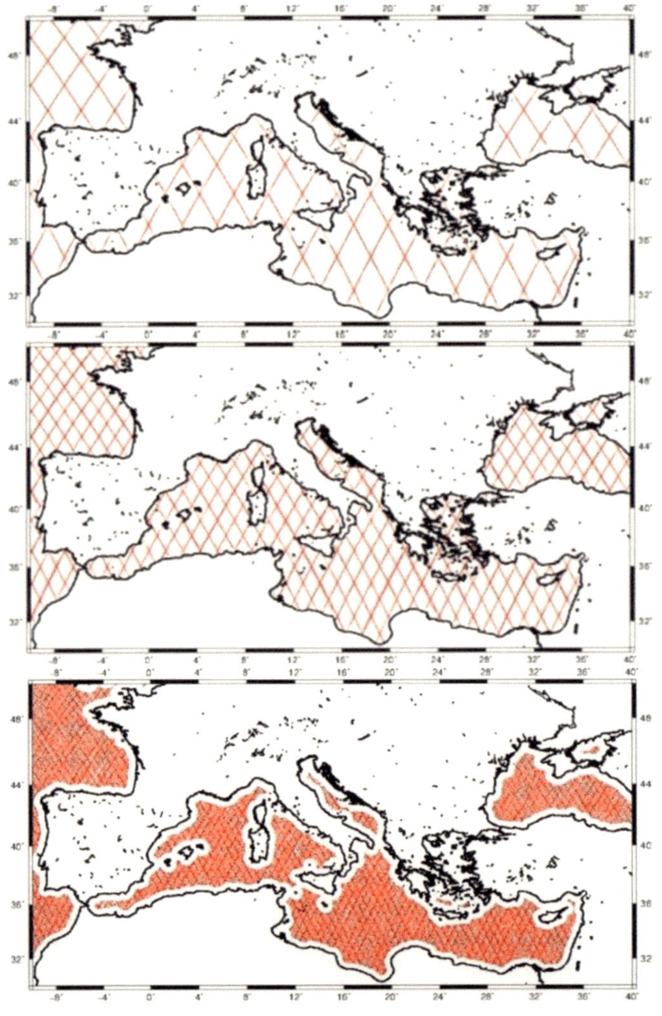

Figure 2.
Jason-1 data distribution. Phase A (up), B (middle) C (bottom).

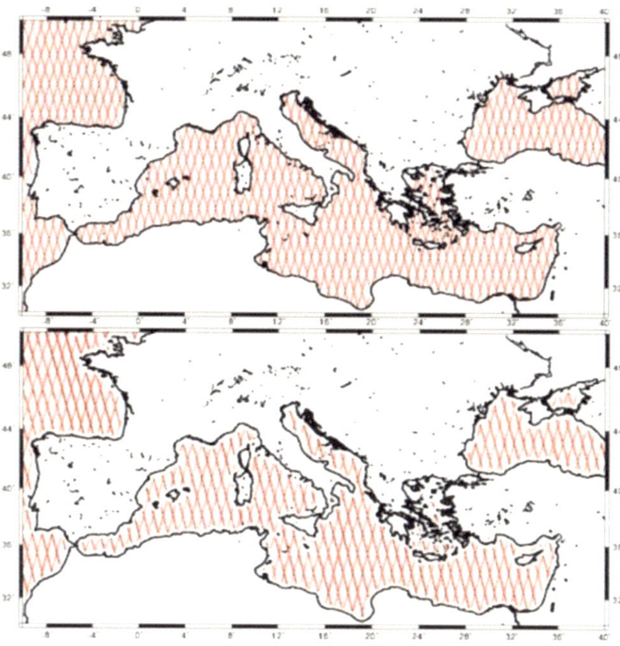

Figure 3.
Envisat data distribution. Phase A (up), B (bottom).

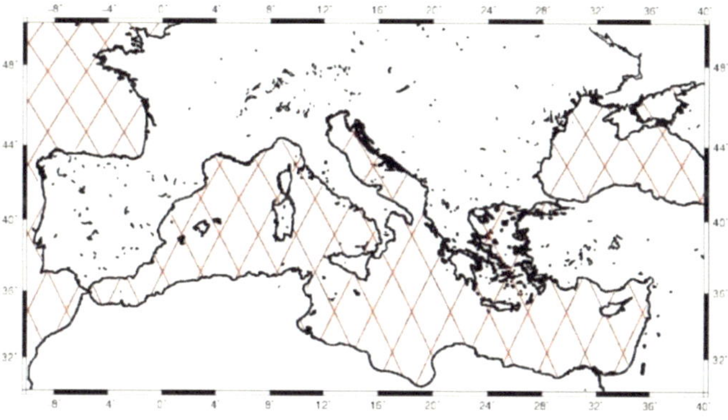

Figure 4.
Jason-2 data distribution.

point values in the Mediterranean Sea (**Figure 4**). Then the second phase of the satellite mission started with a new orbit until May 2017. This second phase of the mission is not included in our study.

The last mission used in this study, is that of the Cryosat-2 altimetry satellite. Cryosat-2 is the successor of Cryosat-1, whose mission was canceled after the failed launch on 2005. Its successor was launched in April 8, 2010 and the data collection started with cycle 4 on July 14 of the same year. Until cycle 73, the satellite records referred to low-resolution (LRM) and synthetic aperture radar (SAR) data, which where then succeeded by SAR and SAR interferometric (SARin) data getting closer to the coastline. Until the end of 2016 78 months of data have been collected with 826,941 point values being available (**Figure 5**).

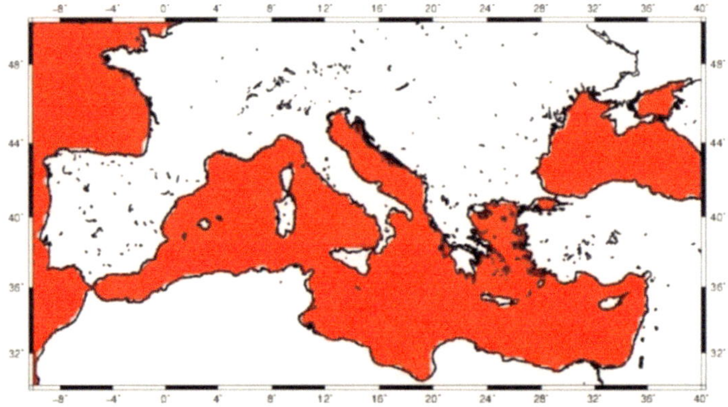

Figure 5.
Cryosat-2 data distribution.

The data for this study were obtained through the Radar Altimeter Database System (RADS) operated by the Delft Institute for Earth-Oriented Space research (DEOS) was used (RADS 2011). RADS provides a platform with a large variety of altimetry missions. This database was chosen as the platform provides not only the geophysical data records, with all the necessary corrections for the altimeter data, but has the additional benefit that all records refer to the same ellipsoid ensuring that there are no reference system problems in the form of biases when combining data from multiple satellite missions [31, 36]. Moreover, the RADS altimetry data are already processed through the method of crossover adjustment for the reduction of radial orbit errors, while it provides a one stop point to collect multi-satellite data instead of retrieving them from various sources.

The platform provides the user various options for the reference surface of SLAs, among which that of the EGM2008 geoid [37], was chosen along with be zero-tide (ZT) geoid model ensuring that the tidal system is the same with the one of the altimetric data. RADS provide also to the user the opportunity to choose along various geophysical corrections and models for the systematic errors that affect the altimeter data quality. In this work the following selections were made:

- ECMWF for the dry tropospheric correction
- MWR (NN) for the wet tropospheric correction
- the smoothed dual-frequency model for the ionospheric correction,
- GOT4.7 for the ocean and pole tide
- the CLS Sea State Bias (SSB) model for the SSB effect.

The Inverse Barometer (IB) correction which is also provided in the geophysical data records, in three different types, local, global and total, was applied to the raw data after the aforementioned geophysical corrections. **Tables 1–4** tabulates the statistics of the SLAs values before and after the application of total IB corrections total inverse barometer corrections. From these Tables, except for the minmax values which are obviously attributed to blunders and their locations are close to shore, it

	nr. values	min	max	mean	std
SLA	689,860	−1.817	0.880	0.007	±0.150
SLA + total IB	689,860	−1.694	0,894	0,059	±0.139

Table 1.
Statistics of JASON-1 phase a data with and without total IB correction. Unit: [m].

	nr. values	min	max	mean	std
SLA	512,995	−1.092	1.069	0.036	±0.160
SLA + total IB	512,995	−0.918	1.150	0.076	±0.148

Table 2.
Statistics of JASON-1 phase B data with and without total IB correction. Unit: [m].

	nr. values	min	max	mean	std
SLA	120,973	−0.749	0.799	0.051	±0.145
SLA + total IB	120,973	−1.136	0.842	0.086	±0.137

Table 3.
Statistics of JASON-1 phase C data with and without total IB correction. Unit: [m].

	nr. values	min	max	mean	std
SLA	881,612	−2.781	1.179	0.032	±0.146
SLA + total IB	881,612	−2.727	1.304	0.078	±0.176

Table 4.
Statistics of ENVISAT data with and without total IB correction. Unit: [m].

can be concluded that IB correction little affect the statistics as difference of few cm (~2–4) are noticed in mean and standard deviation values.

In order to remove the blunders noticed in the SLA data, a 3σ test was performed. The small mean values, close to a zero mean, is a good indication, that all altimetry data used in this work, as to being unbiased (0.7 cm (Jason-1 ph. A) to 5.1 cm (Jason-1 ph. C). The result of the 3σ test is that all SLA values larger than the one of 3σ, are extracted from the final data used. **Table 5** above tabulates the statistics of all SLAs after the 3σ removal test (see the top row of **Tables 1–4** and **6** for comparison). In case of Jason-1, only 26,581 (~2%) of the point values were extracted from three phases, resulting in a reduction of 70 cm. ~1.1% of ENVISAT point values we removed, resulting again in a significant discount at the range of data (~1.4 m). For Jason-2 data, 26,851 point values were removed (~2.8%), while for Cryosat-216,294

	nr. values	min	max	mean	std
SLA	944,941	−1.864	1.686	0.024	±0.156
SLA + total IB	944,941	−0.441	0.441	0.061	±0.134

Table 5.
Statistics of Cryosat-2 data with and without total IB correction. Unit: [m].

	nr. values	min	max	mean	std
SLA	1,061,379	−0.783	1.168	0.047	±0.153
SLA + total IB	1,061,379	−0.787	1.214	0.094	±0.141

Table 6.
Statistics of Jason-2 data with and without total IB correction. Unit: [m].

	nr. values	min	max	mean	std
Jason-1 ph. A	675,041	−0.416	0.416	0.050	±0.124
Jason-1 ph. B	504,763	−0.444	0.444	0.067	±0.141
Jason-1 ph. C	117,443	−0.351	0.351	0.073	±0.117
ENVISAT	871,584	−0.527	0.527	0.072	±0.147
Jason-2	1,031,247	−0.423	0.423	0.082	±0.123
Cryosat-2	928,647	−0.441	0.441	0.061	±0.134

Table 7.
Statistics of all SLAs after the 3σ. Unit: [m].

(~1.7%) of the data were identified as blunders. The reduction of the data range was similar to the other missions from ~2.2 m to 0.8 m (**Table 7**).

3. Sea level anomaly variations in the Mediterranean Sea and correlation with global and regional climatic phenomena

As already mentioned, the present study is focused on the entire Mediterranean basin, within the region bounded between $30° \leq \varphi \leq 50°$ and $-10° \leq \lambda \leq 40°$. In this region, the statistical characteristics of the SLA have been studied using altimetric observations from various satellites for the period 2002–2016. For each mission, the analysis presented here refers to monthly data as only data falling in the specific time period have been used. Each test refers to the use of the entire set of satellite passes for the Mediterranean Sea, so that the SLA variability will be studied in cross-track (2D) direction (see **Figures 2–5**). The computation of empirical covariance functions allowed the investigation of the statistical characteristics of the SLA. The equation of the empirical covariance function for a group of data, for our study (h^{SLA}), under a known spherical distance ψ is [38]:

$$C\left(h_i^{SLA},,h_j^{SLA},,\psi\right) = M\left\{h_i^{SLA},h_j^{SLA}\right\}_{\psi} \tag{1}$$

where, M denotes the mean value operator and i, j the SLA observations at two points in the area under study with a distance ψ. Employing Eq. 1, the empirical covariance functions have been estimated for all available satellite cycles. Given the monthly availability of data, it is implied that for each year 12 covariance functions have been determined. As a result, an analysis of the SLAs variances calculated through the covariances functions has been performed for the whole period that the satellite data cover in this study.

As already mentioned in the introduction of this work, sea level variations can also attributed to climate episodes. Thus, three climate indexes have been studied. Southern Oscillation Index (SOI) reflects the sea response to El Niño/La Niña-Southern Oscillation (ENSO) events indicating the evolution and the volume of the two events in the Pacific Ocean. Positive values may lead to La Niña event while on the contrary, negative values are probably a result of El Niño phenomena. For the computation of the value, differences of pressure between Tahiti and Darwin are taken into account [39–41].

The second index to be investigated was NAO, an index that provides information for climate variations in the North Atlantic Ocean. In the area of interest, positive values indicate dry winters while for the same season warmer and more wet conditions can be found in other part of Europe. On the other hand, negative values indicate humid atmosphere in the south part of Europe and frozen air in the north. For the computation of index values, surface sea-level pressure difference between the Subtropical High and the Subpolar Low are used [42–46]. The most proper index for studying the correlation between the sea level and climatic phenomena is MOI as it refers to pressure differences between Algiers and Cairo or Gibraltar and Israel. In both cases, positive values indicate dry conditions in the Mediterranean mainly in the north-west segment while negative values are connected with cyclones and wet conditions in the west [47–50]. For the present study, data for these indices have been acquired from the Climate Research Unit of the University of East Anglia (http://www.cru.uea.ac.uk/).

Figures 6–9 below depict the SLA variances along with the SOI, NAO and MOI indices for all consecutive months within each year between 2002 and 2011 for Jason-1 satellite, 2008 and 2016 for Jason-2, 2002 and 2012 for ENVISAT satellite and 2010 and 2016 for Cryosat-2 satellite. Consecutive negative values of SOI lower than −0.7 indicate El Niño phenomena and positive values larger than 0.7 La NiñaFor all missions, it can be summarized that despite the delay of ~one semester, there is a connection between the El Niño and the La Niña events and SLA change in the Mediterranean. The smallest value of SOI in early 2005 resulted in large variance in the summer of 2005 both on Envisat (~320 cm^2) and Jason-1(~240 cm^2). Moreover, the evolution of El Niño in Spring of next year had a faster stamp on SLA variation as the values are rising during Summer and Autumn (from 170 cm^2 to 280 cm^2 for Envisat and from 100 cm^2 to 250 cm^2 for Jason-1) [51, 52]. Similar results are found while studying the severe La Niña episodes during the last months of 2007, the first and the last of 2008 and the start of 2009. Strong La Niña during 2010–2011 and El Niño during 2015–2016 resulted in significant variations in SLAs. For Jason-2, the variance from 324 cm^2 in April 2010 decreased to 98 cm^2 while after El Niño the variance from 102 cm^2 in September 2014 reached the 300 cm^2 in February 2015 [53, 54].

The geometry and the shape of the Mediterranean along with the location of ENSO events result that this index is not the most proper for studying the response of SLA to climatic phenomena. Once the NAO is examined, the heavier relation between SLA and pressure can be noticed. The increasing of the index results in fast falls of SLA variation. This is noticed mainly during winter while during summer the variation in sea level due to changes in NAO values are not so instant. Same results are found in similar study [48] which signals that atmospheric forcing is not the contributing factor to the steric sea level variations in the Mediterranean during the summer period. Positive values of the index in early 2007, 2008, and 2011 resulted in a decrease of the variance, from 286 cm^2 in December 2006 to 120 cm^2 in July 2007,

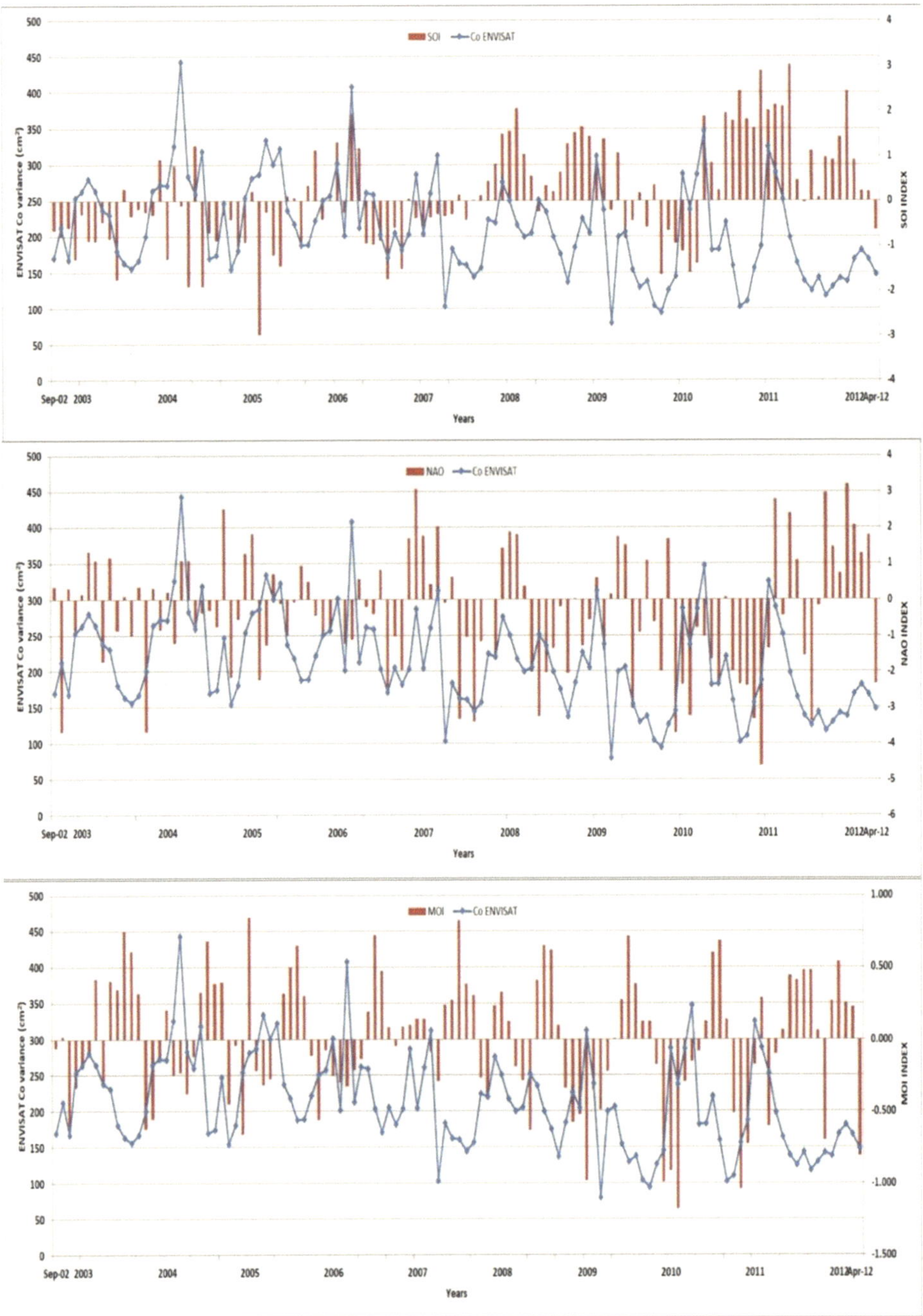

Figure 6.
ENVISAT SLA variances fluctuations from 2002 to 2012 and correlation with SOI (up), NAO (middle) and MOI (bottom).

and from 212 cm^2 in March 2008 to 93 cm^2 in July 2008 for Envisat, while negative values mainly in summer and autumn months increased the variance from 87 cm^2 in September 2002 to 252 cm^2 in February 2003 and from 130 cm^2 in September 2010 reached the 333 cm^2 in early 2011 for Jason-1 [42, 55, 56].

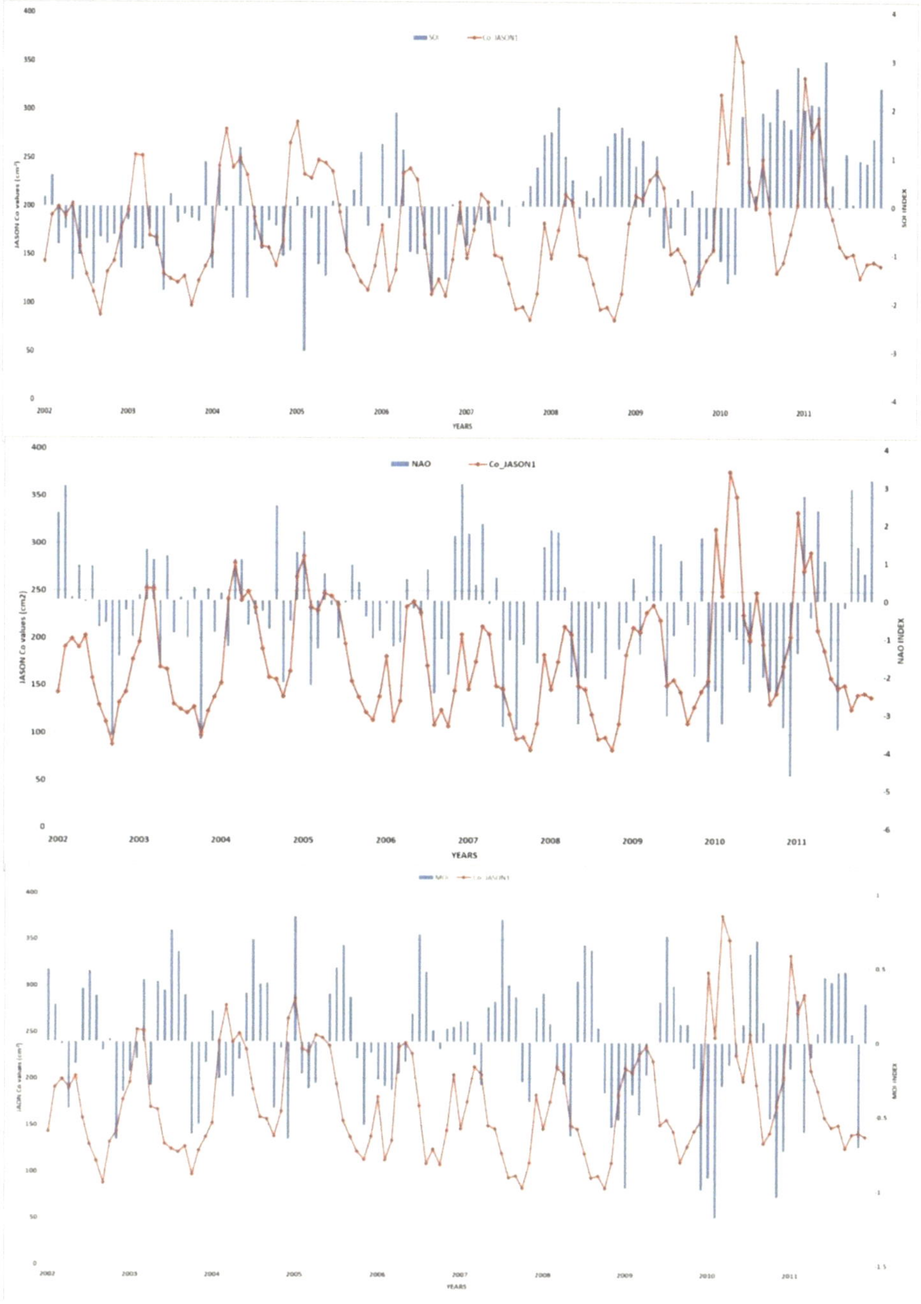

Figure 7.
Jason-1 SLA variances fluctuations from 2002 to 2011 and correlation with SOI (up), NAO (middle) and MOI (bottom).

To assess that, the MOI index has been investigated as well, since it should be the most proper measure of atmospheric forcing contribution to sea level variations in the Mediterranean. As it is clearly depicted in the bottom of all figures above where the SLA variances fluctuations and correlation with MOI is studied, this index is

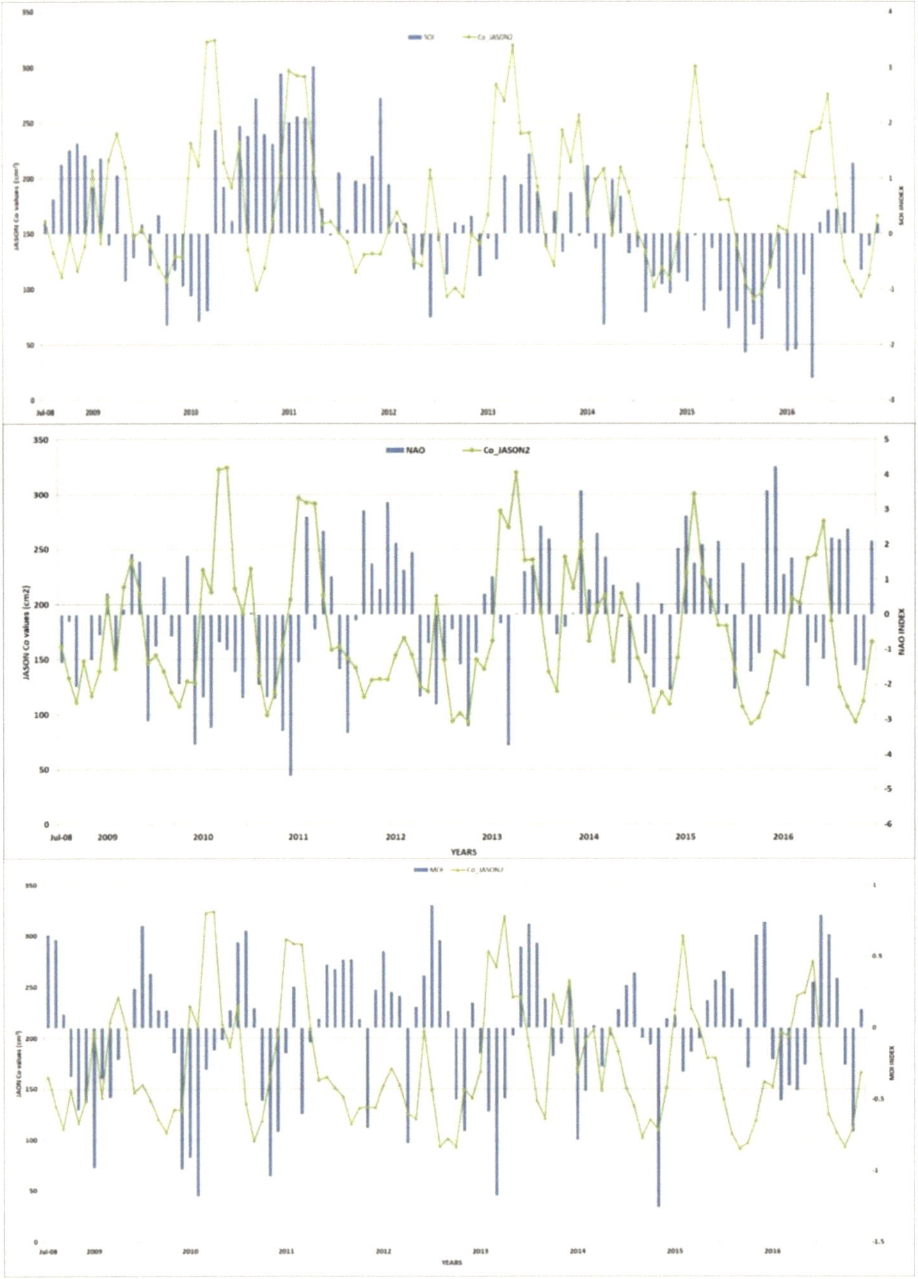

Figure 8.
Jason-2 SLA variances fluctuations from 2008 to 2017 and correlation with SOI (up), NAO (middle) and MOI (bottom).

strongly correlated with SLA values. Consecutive large values of MOI indicate high temperatures and decrease of SLA variances while negative values of the index are connected with SLA rise. These findings are noticed in all missions, for the first case i.e. summer 2004 for Envisat (from 247 cm^2 to 153 cm^2) and Jason-1 (from 192 cm^2 to 114 cm^2), summer 2011 and 2012 for Jason-2 (~110 cm^2 and ~ 80 cm^2) and Cryosat-2

Figure 9.
Cryosat-2 SLA variances fluctuations from 2008 to 2017 and correlation with SOI (up), NAO (middle) and MOI (bottom).

(~120 cm^2 and ~ 110 cm^2) or for negative values in early 2009 (from 138 cm^2 to 243 cm^2), 2010 (from 152 cm^2 to 334 cm^2) for Jason-2 and in late 2012 and early 2013 for Cryosat-2 when the variance increased from 182 cm^2 to 321 cm^2. However, there are some incidents that SLA fluctuations are not connected with MOI variances indicating a stronger correlation with NAO for the same period, signing that currents in Atlantic are also connected with variances in sea level of Mediterranean.

3.1 EOF analysis

EOF analysis is a very used technique in geophysical sciences in order to study any possible spatial modes of variability and how they change with time. Monthly gridded values of SLAs have been used to estimate the principal components of the time series. SLA values have been gridded to a 5 arcmin step and 172 consecutive monthly data were available from September 2002 to December 2016. Note that whenever multi-mission data have been available for a specific month, then the grid generated employs all available information.

The increasing trend that is noticed worldwide ~2 mm/year [18, 57] is also depicted in the Mediterranean (see **Figure 10**) in smaller or larger scales varying through the time or the region under study. Positive trends of 0.3 ± 0.4 mm/year in the Western part and 1.3 ± 0.4 mm/year in the Eastern part [19], 4.54 ± 0.3 mm/year for the coastal areas and 4.28 ± 0.3 mm/year for the open sea were found using altimetry from TOPEX/Poseidon and tides-gauge data [58] and 2.44 ± 0.5 mm/year with similar data spanning 20 years [59]. As it also depicted in **Figure 10**, the sea level trend differs but gradually increases from the Western to the Eastern part. More significant negative trends can be noticed in the Alboran Sea between Spain and Marocco (the lowest ~ – 10 mm/year) and northern to Egypt while the largest

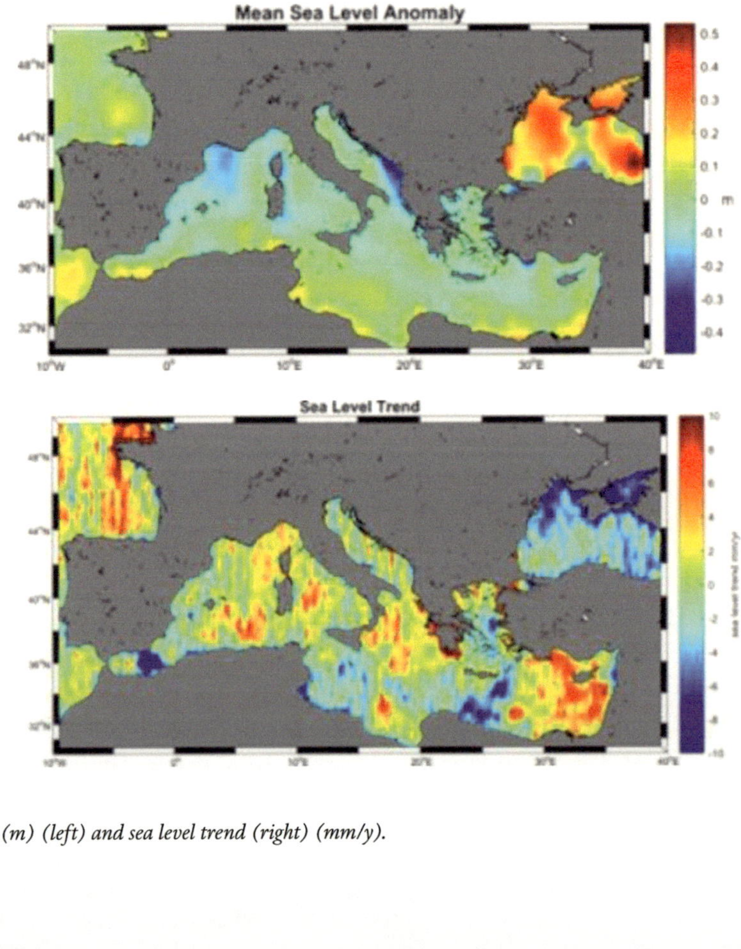

Figure 10.
Mean SLA (m) (left) and sea level trend (right) (mm/y).

positive trends are noticed in areas close to Peloponnese and the Levantine Sea (up to 10 cm/year). All aforementioned areas are connected with strong gyres that affect the sea level (Shikmona Gyre and Asia Minor Current in Levantine Sea, Western and Eastern Alboran Gyre in Alboran Sea) [59–64]. Similar findings are connected with the negative trends in the Ionian Sea [63, 65] and in the North Eastern part of Crete island (Ierapetra Gyre) [66, 67].

Figure 11 depicts the first EOF and first PC before removing the seasonal variability. From this Figure the annual and seasonal pattern in the SLAs are evident with the largest values occurring during the summer months and the smallest ones in the Fall. Moreover, the dominant increasing trend in the SLA is clearly depicted. After removing the trend and the seasonal signal, the EOF analysis was applied to the SLA time series for all data, in order to extract individual dominant modes of the data variability. The monthly SLA field is separated into spatial structures, the empirical orthogonal functions (EOF) and their amplitudes in time, the principle components, which are depicted in **Figure 12**. The annual signal, being 51.2% of the total variability, presents the increase of sea level that is dominant in the whole area. Modes 3 to 6,

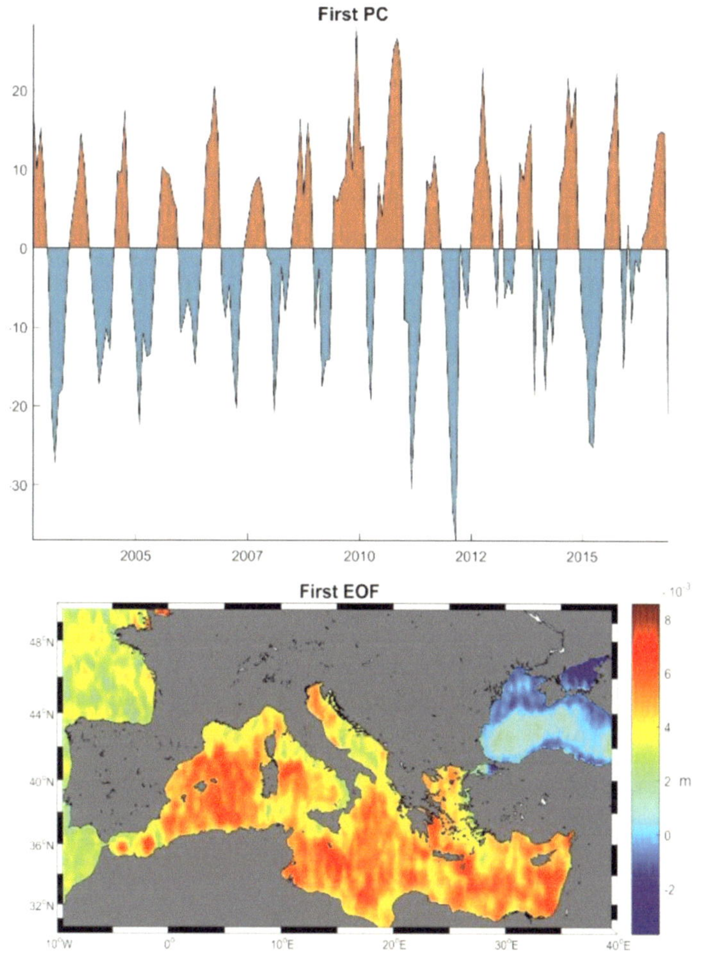

Figure 11.
First PC of SLA and first EOF before removing seasonal variability.

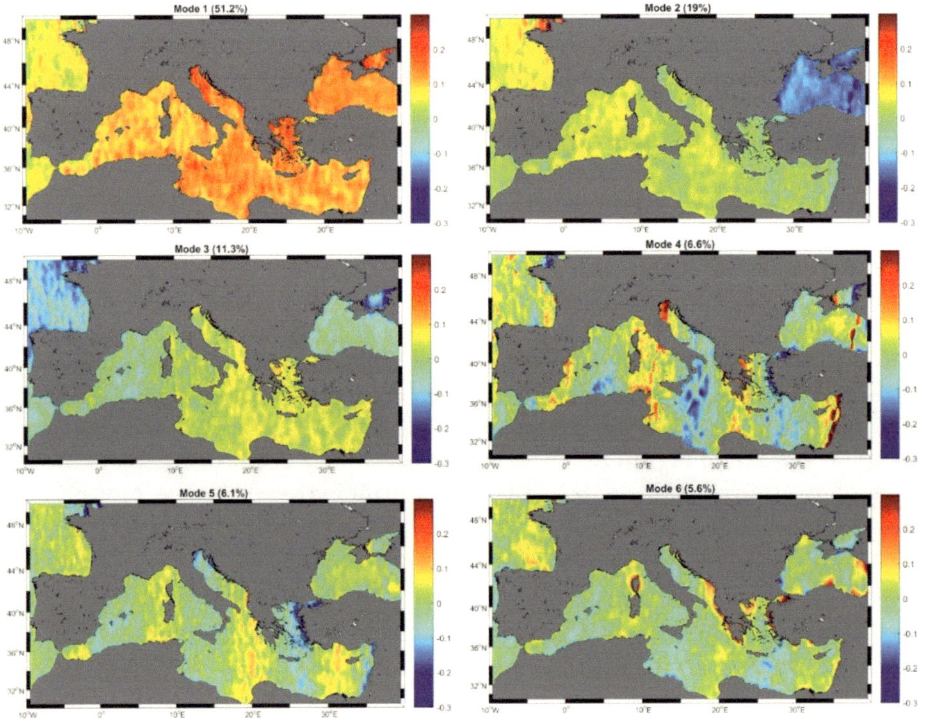

Figure 12.
First six EOF after removing seasonal variability (m).

despite having smaller percentages clearly depict local trends that can be attributed to various characteristics (different values of atmospheric pressure, entrance of ocean water from Atlantic Ocean, lake and river water that change salinity, etc.).

4. Conclusions

An analytical outline of the use of satellite altimetry data from the exact repeat missions of Envisat, Jason-1, Jason-2, and Cryosat-2 to monitor SLA variations has been presented. Through the empirical covariance functions of SLAs, it was found an important annual variation which is obvious for the whole period. This annual change follows the temperature. As the temperature rise, the water gets warmer and expands while the sea contracts when temperature decreases. This cyclo-stationary is further connected to changes in pressure in the sea level as it was clearly depicted in this work.

When the fluctuation of sea level is compared to meteorological phenomena, the El Niño and La Niña result in a slower change in the Mediterranean depending on the intensity of the events. NAO index is stronger connected to SLA variations during the late and early months of the year while MOI, as it refers to variations in pressure in Mediterranean is the most proper measure of atmospheric forcing contribution to sea level variations. Consecutive large values are connected with SLA rise while negative phases indicate decrease in variances of SLA. Finally, when the NAO and MOI are examined together, a correlation exists mainly during the cold months however this finding is not observed during the warm months indicating that circulation in Atlantic little affects the sea level in the Mediterranean.

In the last step in this work, the method of Principal Component Analysis (PCA) was applied to the SLA time series in order to extract individual dominant modes of the data variability. The dominant first PC with 51.2% presents the increase of sea level that is dominant in the whole area. Modes 3 to 6, despite having smaller percentages depict clearly local trends in the sea level in the Mediterranean.

Author details

Dimitrios A. Natsiopoulos[1], Eleni A. Tzanou[2] and Georgios S. Vergos[1*]

1 Laboratory of Gravity Field Research and Applications (GravLab), Department of Geodesy and Surveying, Aristotle University of Thessaloniki, Greece

2 Department of Surveying and Geoinformatics Engineering, International Hellenic University (IHU), Greece

*Address all correspondence to: vergos@topo.auth.gr

References

[1] Fenoglio-Marc L, Tel E. Coastal and global sea level change. Journal of Geodynamics. 2010;**49**(3-4):151-160

[2] Tsimplis MN, Shaw AGP, Pascual A, Marcos M, Pasaric M, Fenoglio-Marc L. Can we reconstruct the 20th century sea level variability in the mediterranean sea on the basis of recent altimetric measurements? In: Remote Sensing of the European Seas. Netherlands: Springer; 2008. pp. 307-318

[3] Gomis D, Tsimplis M, Marcos M, Fenoglio-Marc L, Pérez B, Raicich F, et al. Mediterranean Sea-level variability and trends. In: The Climate of the Mediterranean Region. Cambridge, UK, and New York NY, USA: Cambridge University Press; 2012. pp. 257-299

[4] Weisse R, Bellafiore D, Menéndez M, Méndez F, Nicholls RJ, Umgiesser G, et al. Changing extreme sea levels along European coasts. Coastal Engineering. 2014;**87**:4-14

[5] Idier D, Paris F, Le CG, Boulahya F, Dumas F. Sea-level rise impacts on the tides of the European shelf. Continental Shelf Research. 2017;**137**(April):56-71. DOI: 10.1016/j.csr.2017.01.007

[6] Zerbini S, Raicich F, Prati CM, Bruni S, Del Conte S, Errico M, et al. Sea-level change in the northern Mediterranean Sea from long-period tide gauge time series. Earth-Science Reviews. 2017;**167**:72-87

[7] Ambrosetti E. Demographic Challenges in the Mediterranean. Eur Inst Mediterr, Mediterranean Yearbook 2020. Barcelona, Spain: IEMED Fundació CIDOB; 2020

[8] Andersen TJ, Svinth S, Pejrup M. Temporal variation of accumulation rates on a natural salt marsh in the 20th century - the impact of sea level rise and increased inundation frequency. Marine Geology. 2011;**279**(1-4):178-187. DOI: 10.1016/j.margeo.2010.10.025

[9] Pickering MD, Wells NC, Horsburgh KJ, Green JAM. The impact of future sea-level rise on the European shelf tides. Continental Shelf Research. 2012;**35**:1-15. DOI: 10.1016/j.csr.2011.11.011

[10] Pelling HE, Mattias Green JA, Ward SL. Modelling tides and sea-level rise: To flood or not to flood. Ocean Modelling. 2013;**63**:21-29. DOI: 10.1016/j.ocemod.2012.12.004

[11] Parker A. Comment to assessing sea level rise costs and adaptation benefits under uncertainty in Greece. Environmental Science & Policy. 2014;**38**:178-179. DOI: 10.1016/j.envsci.2013.12.003

[12] Dawson D, Shaw J, Roland GW. Sea-level rise impacts on transport infrastructure: The notorious case of the coastal railway line at Dawlish. England Journal of Transport Geography. 2016;**51**(February):97-109. DOI: 10.1016/j.jtrangeo.2015.11.009

[13] Kuhfuss L, Rey-Valette H, Sourisseau E, Heurtefeux H, Rufray X. Evaluating the impacts of sea level rise on coastal wetlands in Languedoc-Roussillon, France. Environmental Sciences Policy. 2016;**59**:26-34. DOI: 10.1016/j.envsci.2016.02.002

[14] Masciopinto C, Liso IS. Assessment of the impact of sea-level rise due to climate change on coastal groundwater discharge. Science of the Total Environment. 2016;**569-570**:672-680. DOI: 10.1016/j.scitotenv.2016.06.183

[15] Antonioli F, Anzidei M, Amorosi A, Lo Presti V, Mastronuzzi G, Deiana G, et al. Sea-level rise and potential drowning of the Italian coastal plains: Flooding risk scenarios for 2100. Quaternary Science Reviews. 2017;**158**:29-43. DOI: 10.1016/j.quascirev.2016.12.021

[16] Bergillos RJ, Rodriguez-Delgado C, Iglesias G. Wave farm impacts on coastal flooding under sea-level rise: A case study in southern Spain. Science of the Total Environment. 2019;**653**:1522-1531. DOI: 10.1016/j.scitotenv.2018.10.422

[17] Vergos GS, Natsiopoulos DA. Ocean remote sensing altimetric satellites in support of sea level anomalies and mean sea surface modeling. In: Perakis K, Moysiadis A, editors. Advances in Geosciences, 32nd European Association of Remote Sensing Laboratories (EARSeL) Symposia. 2013. pp. 398-423

[18] Miller L, Douglas BC. Mass and volume contributions to twentieth-century global sea level rise. Nature. 2004;**428**(6981):406-409

[19] Tsimplis MN, Álvarez-Fanjul E, Gomis D, Fenoglio-Marc L, Pérez B. Mediterranean Sea level trends: Atmospheric pressure and wind contribution. Geophysical Research Letters. 2005;**32**(20):1-4. DOI: 10.1029/2005GL023867

[20] Tsimplis M, Spada G, Marcos M, Flemming N. Multi-Decadal Sea level trends and land movements in the Mediterranean Sea with estimates of factors perturbing tide gauge data and cumulative uncertainties. Global and Planetary Change. 2010;**76**:63-76

[21] Mahdi H, Hebib T. Mediterranean Sea level trends from long-period tide gauge time series. Acta Oceanologica Sinica. 2020;**39**(1):157-165

[22] Vignudelli S, Cipollini P, Roblou L, Lyard F, Gasparini GP, Manzella G, et al. Improved satellite altimetry in coastal systems: Case study of the Corsica Channel (Mediterranean Sea). Geophysical Research Letters. 2005;**32**(7):1-5

[23] Criado-Aldeanueva F, Del Río VJ, García-Lafuente J. Steric and mass-induced Mediterranean Sea level trends from 14 years of altimetry data. Global Planet Change. 2008;**60**(3-4):563-575

[24] Birol F, Delebecque C. Using high sampling rate (10/20Hz) altimeter data for the observation of coastal surface currents: A case study over the northwestern Mediterranean Sea. Journal of Marine Systems. 2014;**129**:318-333

[25] Grgić M, Nerem RS, Bašić T. Absolute sea level surface modeling for the Mediterranean from satellite altimeter and tide gauge measurements. Marine Geodesy. 2017;**40**(4):239-258. DOI: 10.1080/01490419.2017.1342726

[26] Stammer D, Cazenave A. Satellite Altimetry over Oceans and Land Surfaces. Boca Raton, FL: CRC Press; 2017. pp. 1-618

[27] Knudsen P, Andersen O, Bingham R. Enhanced Mean Dynamic Topography and Ocean Circulation Estimation Using GOCE Preliminary Models. AGU Fall Meet Abstr. Noordwijk, Netherlands: European Space Agency; 2010

[28] Andersen OB. THE DTU12MDT Global Mean Dynamic Topography And Ocean Circulation Model. 2013 Available from: http://earth.esa.int/gut/ [Accesed: April 9, 2020]

[29] Arabelos D, Tziavos IN. Combination of ERS-1 and TOPEX altimetry for precise geoid and gravity recovery in the Mediterranean Sea. Geophysical Journal International. 1996;**125**(1):285-302

[30] Vergos GS, Tziavos IN, Andritsanos VD. On the determination of marine geoid models by least-squares collocation and spectral methods using heterogeneous data. International Association of Geodesy Symposium. 2005;**128**:332-337

[31] Tziavos IN, Vergos GS, Kotzev V, Pashova L. Mean sea level and sea level variation studies in the black sea and the Aegean. International Association of Geodesy Symposium. 2005;**129**:254-259

[32] Vergos GS, Tziavos IN, Andritsanos VD. Gravity data base generation and geoid iviodel estimation using heterogeneous data. International Association of Geodesy Symposium. 2005;**129**:155-160

[33] Cazenave A, Nerem RS. Present-Day Sea level change: Observations and causes. Reviews of Geophysics. 2004;**42**(3):1-20

[34] Church JA, White NJ, Konikow LF, Domingues CM, Cogley JG, Rignot E, et al. Revisiting the Earth's sea-level and energy budgets from 1961 to 2008. Geophysical Research Letters. 2011;**38**(18):n/a-n/a. DOI: 10.1029/2011GL048794

[35] Nerem RS, Leuliette É, Cazenave A. Present-Day Sea-level change: A review. Comptes Rendus - Geoscience. 2006;**338**(14-15):1077-1083

[36] Fernandes MJ, Barbosa S, Lázaro C. Impact of altimeter data processing on sea level studies. Sensors. 2006;**6**(3):131-163

[37] Pavlis NK, Holmes SA, Kenyon SC, Factor JK. The development and evaluation of the earth gravitational model 2008 (EGM2008). Journal of Geophysical Research. 2012;**117**:B04406. DOI: 10.1029/2011JB008916

[38] Heiskanen M. Physical Geodesy. W. A. Heiskanen and H. Moritz. pp. ix + 364, and numerous diagrams. W. H. Freeman and Co., San Francisco and London, Price 100 s. Geol Mag. 1967;**104**(3):302

[39] Allan RJ, Nicholls N, Jones PD, Butterworth IJ. A further extension of the Tahiti-Darwin SOI, early ENSO events and Darwin pressure. Journal of Climate. 1991;**4**:743-749

[40] Können G, Jones P, Kaltofen M, Allan R. Pre1866 extensions of the southern oscillation index using early Indonesian and Tahitian meteorological readings. Journal of Climate. 1998;**11**:2325-2339

[41] Ropelewski CF, Jones PD. An extension of the Tahiti–Darwin southern oscillation index. Monthly Weather Review. 1987. doi;**115**:2161-2165. DOI: 101175/1520-0493(1987)115<2161: AEOTTS>20CO;2

[42] Tsimplis MN, Josey SA. Forcing of the Mediterranean Sea by atmospheric oscillations over the North Atlantic. Geophysical Research Letters. 2001;**28**(5):803-806

[43] Osborn TJ. Recent variations in the winter North Atlantic oscillation. Weather. 2006;**61**(12):353-355. DOI: 10.1256/wea.190.06

[44] Osborn TJ. Winter 2009/2010 temperatures and a record-breaking North Atlantic oscillation index. Weather. 2011;**66**(1):19-21. DOI: 10.1002/wea.660

[45] Wakelin SL, Woodworth PL, Flather RA, Williams JA. Sea-level dependence on the NAO over the NW European continental shelf. Geophysical Research Letters. 2003;**30**(7):1403. DOI: 10.1029/2003GL017041

[46] Woolf DK, Shaw AGP, Tsimplis MN. The influence of the North Atlantic oscillation on sea-level variability in the North Atlantic region. The Global Atmosphere and Ocean System. 2003;**9**(4):145-167

[47] Palutikof J. Analysis of Mediterranean climate data: Measured and modelled. In: Mediterranean Climate. Berlin Heidelberg: Springer; 2003. pp. 125-132

[48] Tsimplis MN, Shaw AGP. The forcing of mean sea level variability around Europe. Global Planet Change. 2008;**63**(2-3):196-202

[49] Supić N, Grbec B, Vilibić I, Ivančić I. Long-term changes in hydrographic conditions in northern Adriatic and its relationship to hydrological and atmospheric processes. Annales de Geophysique. 2004;**22**(3):733-745

[50] Sušelj K, Bergant K. Mediterranean Oscillation Index. Vol. 8. Geophysical Research Abstracts; 2006. p. 02145. SRef-ID: 1607-7962/gra/EGU06-A-02145

[51] Brönnimann S. Impact of El Niño-southern oscillation on European climate. Reviews of Geophysics. 2007;**45**:RG3003. DOI: 10.1029/2006RG000199

[52] Huang B, Thorne PW, Banzon VF, Boyer T, Chepurin G, Lawrimore JH, et al. Extended reconstructed sea surface temperature, version 5 (ERSSTv5): Upgrades, validations, and intercomparisons. Journal of Climate. 2017;**30**(20):8179-8205

[53] Zhang RH, Zheng F, Zhu J, Wang Z. A successful real-time forecast of the 2010-11 la Niña event. Scientific Reports. 2013;**3**(1):1-7

[54] Paek H, Yu JY, Qian C. Why were the 2015/2016 and 1997/1998 extreme El Niños different? Geophysical Research Letters. 2017;**44**(4):1848-1856

[55] Josey SA, Somot S, Tsimplis M. Impacts of atmospheric modes of variability on Mediterranean Sea surface heat exchange. Journal of Geophys Research and Ocean. 2011;**116**(2):C02032. DOI: 10.1029/2010JC006685

[56] Tsimplis MN, Calafat FM, Marcos M, Jordà G, Gomis D, Fenoglio-Marc L, et al. The effect of the NAO on sea level and on mass changes in the Mediterranean Sea. Journal of Geophys Research and Ocean. 2013;**118**(2):944-952. DOI: 10.1002/jgrc.20078

[57] Wöppelmann G, Letetrel C, Santamaria A, Bouin MN, Collilieux X, Altamimi Z, et al. Rates of sea-level change over the past century in a geocentric reference frame. Geophysical Research Letters. 2009;**36**(12):1-6

[58] Mangiarotti S. Coastal Sea level trends from TOPEX-Poseidon satellite altimetry and tide gauge data in the Mediterranean Sea during the 1990s. Geophysical Journal International. 2007;**170**(1):132-144. DOI: 10.1111/j.1365-246X.2007.03424.x

[59] Bonaduce A, Pinardi N, Oddo P, Spada G, Larnicol G. Sea-level variability in the Mediterranean Sea from altimetry and tide gauges. Climate Dynamics. 2016;**47**:2851-2866

[60] Stanev EV, Le Traon PY, Peneva EL. Sea level variations and their dependency on meteorological and hydrological forcing: Analysis of altimeter and surface data for the black sea. Journal of Geophys Research and Ocean. 2000;**105**(C7):17203-17216

[61] Rio MH, Pascual A, Poulain PM, Menna M, Barceló B, Tintoré J. Computation of a new mean dynamic

topography for the Mediterranean Sea from model outputs, altimeter measurements and oceanographic in situ data. Ocean Science. 2014;**10**(4):731-744

[62] Rio MH, Poulain PM, Pascual A, Mauri E, Larnicol G, Santoleri R. A mean dynamic topography of the Mediterranean Sea computed from altimetric data, in-situ measurements and a general circulation model. Journal of Marine Systems. 2007;**65**(1-4 SPEC. ISS):484-508

[63] Pinardi N, Zavatarelli M, Adani M, Coppini G, Fratianni C, Oddo P, et al. Mediterranean Sea large-scale low-frequency ocean variability and water mass formation rates from 1987 to 2007: A retrospective analysis. Progress in Oceanography. 2015;**132**:318-332. DOI: 10.1016/j.pocean.2013.11.003

[64] Mohamed B, Abdallah AM, Alam El-Din K, Nagy H, Shaltout M. Inter-annual variability and trends of sea level and sea surface temperature in the Mediterranean Sea over the last 25 years. Pure and Applied Geophysics. 2019;**176**(8):3787-3810

[65] Gačić M, Civitarese G, Eusebi Borzelli GL, Kovačević V, Poulain PM, Theocharis A, et al. On the relationship between the decadal oscillations of the northern Ionian Sea and the salinity distributions in the eastern Mediterranean. Journal of Geophysical Research. 2011;**116**(C12):C12002. DOI: 10.1029/2011JC007280

[66] Fusco G, Manzella GMR, Cruzado A, Gacic M, Gasparini GP, Kovacevic V, et al. Variability of mesoscale features in the Mediterranean Sea from XBT data analysis to cite this version: Variability of mesoscale features in the Mediterranean Sea from XBT data analysis. Annales de Geophysique. 2003;**21**:21-32

[67] Marullo S, Napolitano E, Santoleri R, Manca B, Evans R. Variability of Rhodes and Ierapetra gyres during Levantine intermediate water experiment: Observations and model results. Journal of Geophys Research. 2003;**108**(C9):8119. DOI: 10.1029/2002JC001393

Chapter 3

Sea-Level Changes

Tarek M. El-Geziry

Abstract

Tide gauge records and satellite altimetry have demonstrated that the sea level is rising on global and relative (regional/local) scales. Globally, the rate of sea-level rise (SLR) in the past two decades is faster than at any time. During the most recent era, 2006–2018, the global SLR rate was 3.7 mm/year, i.e. nearly three times faster than during 1901–1971 (1.3 mm/year). This is mainly attributed to two main reasons: (1) seawater thermal expansion due to climate change and global warming, and (2) ice melting of the Arctic and Antarctic regions. Additionally, the vertical land movement (subsidence/rise) can impact the calculated relative SLR rates. SLR is projected to continue if global warming will continue. SLR has a destructive impact on coastal cities, especially coastal low-lying areas. Factually, it is not only human infrastructures that are at risk from the SLR and coastal flooding, but also coastal environments such as coastal wetlands, seagrass beds, rocky shores, and sandy beaches are vulnerable to such a rise and flooding. This chapter aims at highlighting the SLR issue on global and relative scales, by using both tide gauges and altimeter tools.

Keywords: climate change, sea-level variations, sea-level rise, relative sea level, tide gauge, altimeter

1. Introduction

The importance of the sea-level rise (SLR) issue stems from its direct impact on human lives, coastal infrastructures and constructions, and the coastal environment not only on local and regional scales but also on a global one. Throughout this chapter, the problem will be discussed depending on previous research results from both global and relative (regional and local) points of view to adequately evaluate the problem. The chapter consists of five sections, including this introductory one, which introduces the terminologies used to study the sea-level variations, in addition to the impact of climate change on the observed sea levels. Section two focuses on the tools for measuring that can be used to measure sea level. Section three discusses the global SLR causes and impacts. Section four introduces the relative SLR problem and its impact from regional and local perspectives. The chapter ends with conclusions followed by the set of references listed in the main manuscript, to which those who are interested may refer to.

1.1 Sea-level variations

The observed sea-level variations rely on a variety of variables, including storm surges, astronomical (tidal) harmonic components, the seasonal cycle, interannual to secular variability, and, finally, variations at geological and interglacial scales [1–3]. At any location and at any time (t), the observed sea level (η) is the sum of three elements: the mean sea level (MSL), the tidal component (X), and the residual component (R). This is mathematically expressed by (Eq. (1)):

$$\eta(t) = MSL + X(t) + R(t) \tag{1}$$

The mean sea level (MSL) is the average relative sea level over a long enough time to average out transients like waves and tides [4]. According to [5], the MSL is usually defined as the average value of the observed hourly level over at least 1 year, ideally more than 19 years, to average over the nodal cycle of 18.61 years in tidal amplitudes and phases and to average out the weather. The MSL series can be identified by specific titles, such as monthly MSL and annual MSL [6].

The tidal component (X(t)) is the coherent component of sea level that reacts to astronomical forcing directly or indirectly. Tides are thus described as the periodic rise and fall of a body of water caused by gravitational interactions between the Sun, Moon, and Earth [6]. In reality, the relative positions of the three celestial bodies cause the most visible variations in the magnitude of tides [7]. Tides can be represented analytically as the finite sum of harmonic constants [8] as shown in (Eq. (2)):

$$X(t) = \sum_{n} A_n cos\left(\frac{2\partial}{T_n}t + \phi_n\right) \tag{2}$$

where A_n is the amplitude of a harmonic component (m), T_n is the period of the specified harmonic component (s), and ϕ_n is the phase of the harmonic component.

The amplitudes and phases of the astronomical harmonic constituents are heavily influenced by the local geography [8, 9].

Approximately 390 tidal constituents were early identified [10], the most important of which are formed by the gravitational attraction between the three celestial bodies: the Sun, Moon, and Earth. The main lunar semidiurnal tide component M_2 is often the largest recognized tidal constituent. This tidal component's tidal-producing force is twice as strong as that of the K_1 tide, the main diurnal constituent [1]. Five constituents are particularly important in modeling applications: K_1, O_1, M_2, S_2, and N_2 [7, 11–13]. This is because these constituents are important for any tidal signal and are adequate to compute variations in tidal levels and currents [9]. Two extreme tidal occurrences are associated with the regular astronomical tides: spring and neap tides. These tides are caused primarily by the combined gravitational influences of the Sun and Moon in relation to their relative positions to each other. The extreme gravitational force between the two celestial entities is extracted when the Moon's course aligns with that of the Sun (new and full Moon phases), resulting in the spring tide. The neap tide, on the other hand, occurs when the Moon's course is normal to the Sun's (1st and 3rd quarter phases). Spring tides have the greatest high tides and lowest low tides, while neap tides have the lowest high tides and highest low tides [7].

The residual (R(t)), also known as surge, is the local shift in ocean elevation along a shore caused by a storm. It is calculated by subtracting the astronomic tidal elevation from the overall elevation and usually lasts a few hours [6]. When wind-generated waves ride on top of the surge, the total instantaneous elevation may be much higher than the predicted surge plus the astronomic tide. This is referred to as a storm surge. Storm surges can be disastrous, particularly on low-lying coastlines. Flooding produced by storm surges has the potential to harm not only coastal structures and human infrastructures but also human lives and the ecosystem along the coast, such as wetlands, seagrass beds, and shorelines. The deltas are one of the most vulnerable areas to storm surges and flooding.

Those who are interested to get more knowledge on the fundamentals and theories of sea-level variations may refer to [1, 5].

1.2 Impact of climate change on observed sea level

Extensive and precise climate monitoring showed unequivocal evidence of recent and accelerating global warming. Climate Change is defined as a change in the condition of the climate that can be identified by changes in the mean and/or variability of its attributes and that lasts for an extended period (decades or longer) [14]. Climate change may be caused by natural processes or by ongoing human-induced modifications in the composition of the atmosphere/land use [14]. The effects of climate change will vary considerably by region on a global scale. For example, warming is expected to be greater over continents than over oceans and to be greatest in the world's polar areas. According to [15], oceans will become increasingly acidic as carbon dioxide is absorbed by marine creatures and combined with water to form carbonic acid. This acidification can harm coral reefs and alter the ecosystems of a variety of fish, shellfish, and other resources on which people rely. The impact of climate change on the observed sea level can be declared in two main terms: the sea-level rise (SLR) and the associated flooding phenomena, e.g. storm surges. Using a combination of satellite altimeter data and conventional measurements of tide gauges, scientists have determined that the sea level is rising worldwide and that the rate of rise is likely to accelerate [16]. The SLR is a significant consequence of climate change, both for societies and the ecosystem. The 20th century warming is very likely to have added considerably to the observed SLR, through the thermal expansion of seawater and widespread loss of land ice [17]. Climate change is expected to decrease the amount of water frozen in glaciers and ice caps due to increased melting and evaporation. Greater melting and evaporation on the Greenland and Antarctic ice sheets are also expected, but this may be offset by higher precipitation [18]. Extreme high water levels, storm surges and coastal flooding will occur with increasing frequency (i.e. with reduced return period) as a result of mean SLR. Their frequency may be further increased if storms become more frequent or severe as a result of climate change [4].

Those who are interested in more details on climate change and the different theories proposed to explain this global phenomenon may refer to [19–21] in addition to the reports published by the Intergovernmental Panel on Climate Change (IPCC), e.g [17, 18].

2. Measurement of sea level

2.1 Tide gauge equipment

A tidal gauge, also known as a sea-level recorder, is a device that measures the change in sea level relative to a specified reference known as a "datum." Sensors

Figure 1.
Tide staff gauge in Alexandria Western Harbor [22].

continually record the water level's height in relation to a height reference surface near the geoid. Tide gauges are split into two main categories:

1. Non-recording tide gauges, also known as staff or pole gauges (**Figure 1**), need an observer to record the level data regularly. A staff gauge is a simple style of tidal gauge used to monitor the sea level. It has a graded vertical board with a width of 150–250 mm and a thickness of 100 mm. Staff gauges of various heights are available, and the appropriate height is determined based on the conditions. The height markers are graduated to a minimum count of 5–10 cm. The staff gauge is mounted vertically at a known elevation. If it is unknown, it should be measured by leveling. The differences in water level are manually monitored by viewing the staff directly from a distance.

2. Recording tide gauges, in which the sea level is recorded by the instrument itself. There are basically four types of these gauges: the floating system gauge, the pressure system gauge, the acoustic system gauge and the radar system gauge.

The floating system gauge, also known as stilling well tide gauge, is likely the most popular of all sea-level measuring techniques on a global scale (**Figure 2**). These gauges were once used at every port and were the principal technique used to collect sea-level data. A well's function is to filter out the wave activity so that tides and longer-period processes can be correctly observed. It is most usually linked with having a float gauge in the well driving a pen and chart recorder or, more recently, a shaft encoder so that sea level height readings can be automatically digitized [23].

Pressure sensor gauges are fixed directly in the sea beneath the sea surface to monitor subsurface pressure. In the old pressure gauges such as the Water Level Recorder 7 (WLR7) of Aanderaa, the recorded data used to be stored in an internal data storage unit (DSU) (**Figure 3**), while in the modern gauges, the sensor is often connected through a cable that carries power and signal lines to an onshore control and logging unit (**Figure 4**).

Figure 2.
Alexandria Western Harbor tide gauge: the still-well (right) and the recording drum sheets (left).

Figure 3.
WLR7 and its DSU (https://www.comm-tec.com/prods/mfgs/Aanderaa/).

The acoustic system gauge depends on an acoustic transducer, which can be positioned vertically above the water surface to perform this type of measurement. However, the sensor of the acoustic gauge is preferably housed inside a tube that offers some surface stilling and safeguards the apparatus so that it can operate continuously and reliably under any circumstances where the reflected signals may be lost.

Lastly, radar tide gauges comprise the technology and software required to transform radar measurements into sea-level height. An example of this radar system is

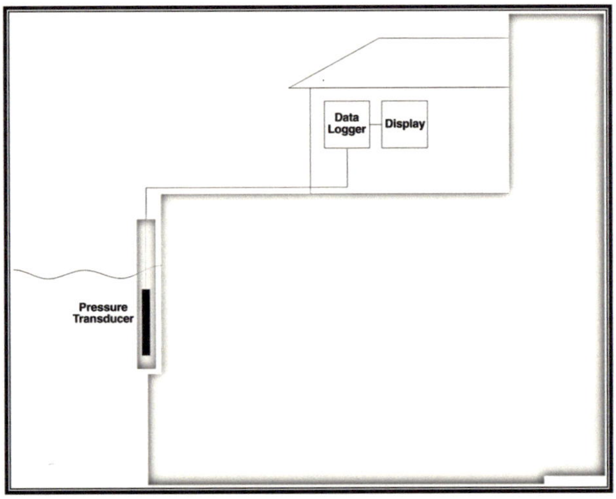

Figure 4.
Modern pressure gauge system [24].

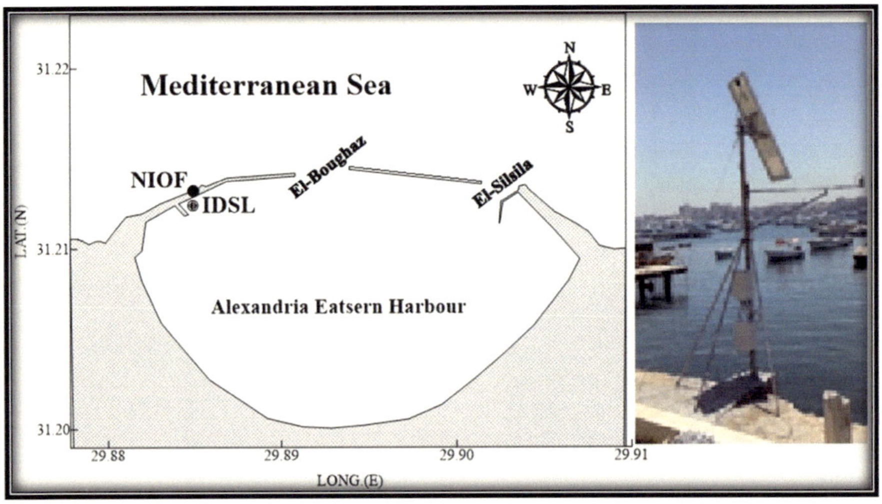

Figure 5.
Radar tide gauge in Alexandria Eastern Harbor [25].

the Inexpensive Device of Sea Level (IDSL) system installed in many Mediterranean harbors, e.g. Alexandria Eastern Harbor in Egypt, as shown in **Figure 5**. The output signals are frequently compatible with current data recorders or can be linked to a communication network. They, like many modern systems, can be configured using a portable computer.

2.2 Altimeter approach

Since the early 1990s, altimeter measurement, the measurement of sea surface height from space, has produced an accurate estimate of changes in sea level every 10 days over the open ocean, attributed to the satellite's frequent sampling capabilities and global coverage [26, 27]. The European Space Agency launched its radar altimeter onboard ERS-1 in 1991. It performed well, but it did not meet the standards for regional or global sea-level change research [28]. Due to land interference within the radar echo in the coastal area, this early altimeter approach built for the open ocean did not produce valid sea-level data within 20 km of the coast [29]. The TOPEX/POSEIDON, which launched in 1992, heralded a new era in satellite altimetry, with the altimeter and orbit errors being only a few centimeters apart, resulting in sea-level observations that were accurate to 3–4 cm [28]. Factually, coastal altimetry has been developed to increase data quality closer to the shore with higher spatial resolution, to extend the satellite-based sea-level record toward the coast with quality comparable to that of the open ocean [30].

Nowadays, coastal flooding, erosion, coastline movement, maritime security, marine pollution, water quality, marine ecology shifts, several marine biophysical features, and atmospheric and oceanic drivers of change have all been effectively monitored using satellite altimetry [31]. The system of sea height altimeter measurement is depicted in **Figure 6**. The main measurement delivered by a satellite altimeter system is the Range (R) [31], which can be calculated using (Eq. (3)):

$$R = \frac{c \times t}{2} \tag{3}$$

where c is the speed of light and t is the travel time of the radar pulse down and up.

In practice, the satellite altitude converts the predicted range R to the instantaneous sea surface level (height) (H_{isl}). Satellite altitude, denoted as (H), is defined as the distance in the normal direction between the satellite center of mass and the reference ellipsoid, as indicated in (Eq. (4)) [22]:

$$H_{isl} = H\text{-}R \tag{4}$$

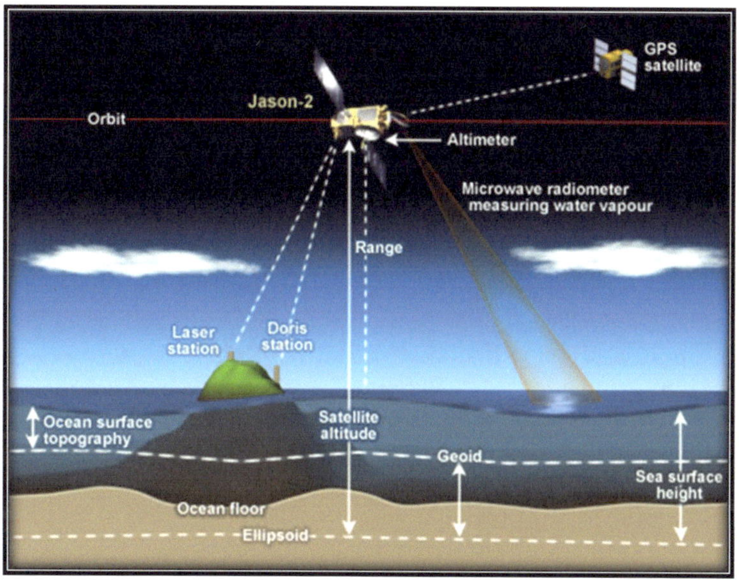

Figure 6.
Altimeter system (https://sealevel.jpl.nasa.gov/missions/technology/).

3. Global sea-level rise (SLR) issue

Measuring sea-level change and understanding its causes has greatly improved in recent years, owing mostly to the availability of new in situ and remote sensing measurements [32]. While the sea level has stayed nearly constant over the last two to three millennia, fluctuations have been recorded since the beginning of the industrial epoch [33]. Global sea level is rising, according to direct data from long-term tide gauges and global satellite altimetry. Tides gauge data showed a global mean SLR of 1.6 to 1.8 mm/year over the 20th century [34]. Between 1993 and 2009, high-precision satellite altimetry implies a recent worldwide acceleration with rates as high as 3.4 mm/year [35]. The IPCC treated the issue of SLR as a universal concern caused by global climate change and greenhouse gas emissions. The IPCC included a chapter in each of its released reports that outlines the updated problem of SLR and sheds insight on the many predictions and scenarios to explore this worldwide issue. In its recent report [18], it is revealed that between 1901 and 2018, the global mean sea level rose by 0.20 m. Between 1901 and 1971, the average rate of sea level rise was 1.3 mm/year, increasing to 1.9 mm/year between 1971 and 2006, and then to 3.7 mm/year between 2006 and 2018. Throughout the period 1993–2019, the altimeter records revealed a global mean SLR rate of 3.3 mm/year [36].

Several natural phenomena contribute to global SLR [33, 37–40]: thermal expansion of seawater due to ocean warming, as well as water mass input from glaciers and ice sheets melting. Over the last century, the global ocean has warmed faster than it has since the end of the previous deglacial transition (about 11,000 years ago) [18]. Between 1992–1999 and 2010–2019, the rate of ice sheet loss accelerated by a factor of 4. Between 2006 and 2018, ice sheets and glacier mass loss were the primary contributors to global mean sea-level rise [18]. Since the early 1990s, various remote sensing tools (airborne and satellite radar and laser altimetry; synthetic aperture radar interferometry (InSAR), and, since 2002, space gravimetry from the Gravity Recovery and Climate Experiment (GRACE) mission) have provided reliable data on the polar ice sheets' mass balance. These findings revealed that mass loss in Greenland and West Antarctica is accelerated [41]. Indeed, ice sheets accounted for less than 15% of global SLR between 1993 and 2003 [4]. However, their contribution has roughly doubled since 2003 [42, 43].

In addition to these natural phenomena, the anthropogenic (man-made) element plays a key influence on the observed SLR. Since at least 1971, human involvement has most certainly been the primary driver of the recent increased global SLR [18].

There is an increasing consensus that an accelerating SLR scenario due to climate warming will have significant impacts on the coastal zone [44]. Changes in the MSL can gradually alter morphological characteristics, pollute subsurface water with salt intrusion, and render coastal areas inhospitable or unsuitable for agriculture [45]. The SLR can have a harmful influence on coastal areas, causing flooding, property damage, and, in some cases, loss of life [45, 46]. Storminess variations may cause additional changes in extremes [45, 47]. Historically, sea-level extremes have increased in lockstep with increases in the MSL in coastal sites. Using this as a foundation, one may relate sea-level extremes to the MSL, allowing one to predict future extremes and return periods [48]. Another way to assess the impact of the SLR is to determine the probabilistic properties of the nontidal residuals (component of storm surge and waves above the tidal variations). Recent studies have focused on the combined consequences of storm intensification, storm surge, and gradual increase in the SLR. On the Atlantic coasts of Europe and Canada, physical factors have been examined using tide/storm surge models [49–52].

Future global SLR projections are complicated due to uncertainty in modeling the many contributory processes, which rely on the understanding of the processes that drive sea level increases as well as trustworthy data to check and calibrate models [40]. Although historical sea level trends are useful for planning for future changes, they are insufficient for estimating risk in the face of future uncertainties [53]. The main components of climate-driven sea-level rise—thermal expansion, glaciers and ice caps, the Greenland ice sheet, and the Antarctic ice sheet—are now projected, though solid ice discharge (SID) from the ice sheets remains difficult to constrain [18]. These projections are frequently created utilizing sets of climate models ranging from simple climate models to intermediate complexity models, comprehensive climate models, and Earth System Models. These models simulate changes depending on a collection of anthropogenic forcing scenarios. According to the IPCC's Fourth Assessment Report (AR4), the global mean sea level would rise by up to 60 cm by 2100 as a result of ocean warming and glacier melting [4]. This forecast increased in the AR5 and AR6 to range between 52 and 98 cm by 2100 for the highest emissions scenario and 28–61 cm for the lowest emissions scenario [18, 54]. To plan for changes due to future sea levels at the local level, local forecasts of SLR that allow varying risk tolerances and cover a variety of periods useful for planning purposes are required [55].

4. Relative (Regional/Local) SLR issue

The global impact of SLR is not uniform. This is attributed to the vertical motion (rise or fall) of the sea surface itself and the vertical motion (rise or subsidence) of the land surface near the sea [38, 56]. The supply of sediments, the wave and current climatology, erosion, and gravitational collapse are also elements that cause variations in the observed coastal sea level. This change in coastal sea level is known as the relative sea-level change, and it can be monitored using a tide gauge at specified coastal locations. Recently, a satellite that detects the motion of the sea surface relative to the center of the Earth (known as a geocentric measurement) has been developed as a sophisticated instrument for assessing the relative SLR issue. A proper assessment of the risks associated with the SLR requires distinguishing between global sea level and relative sea level.

Relative sea-level change is, in fact, the most essential metric for measuring the effects of SLR on infrastructure, property, and ecosystems. Local subsidence of the land causes relative SLR to be greater than global SLR in many regions of the world, especially around several large towns built on deltas. Relative SLR increases the frequency and severity of coastal flooding in low-lying locations, as well as coastal erosion along most sandy coasts [18]. Also, deltas, estuaries, barrier islands, and coral reef communities are among the most vulnerable environments to SLR.

Local changes in the ocean temperature and salinity fields can cause local sea-level changes via variations in the density and volume of the water columns (thermosteric and halosteric effects, respectively) [37, 57–60]. As a result, they cause geographical variation in the rates of sea-level rise. Land discharge fluxes may also affect the observed relative sea level. The input of freshwater from land into the ocean alters the density structure and thus the ocean circulation [61, 62]. This causes regional dynamical alterations in sea level on time intervals ranging from interannual to multidecadal [37, 61–63]. Furthermore, the movement of water mass from land to ocean induces an elastic response of the solid Earth, which deforms ocean basins and coastal morphometry, affecting the observed local sea level [37].

Tide gauge records were used at different coastal regions to calculate the relative SLR rates. The relative sea-level rise rate in the Mediterranean basin was 1.1–1.3 mm/year for the 20th century [64], accelerated to 3.4 mm/yr. in its northwest region for the period 1990 to 2009 [65] and 2.4 ± 0.5 mm/yr. for the period 1993 to 2012 [66]. Shirman [67] observed a SLR of 10 mm/yr. off the Israeli coast in the Levantine Basin between 1958 and 2001. The same rate was found between 1992 and 2002 [68]. Relative SLR rates of 10.6 mm/yr. and 9 mm/yr., respectively, in the Ionian Sea and the Adriatic Sea throughout the decade of 1990–2000 were calculated [69]. The eastern Mediterranean basin had a SLR rate of 0.11 m/110 years, i.e. 1.1 mm/yr [70]. Using the Argo data from 2004 to 2008 [71], investigated the steric sea-level variations in the eastern Mediterranean basin. The results revealed that while in the Ionian basin the total steric sea level change is characterized by strong annual variations (amplitude: 5.9 cm) and a positive trend of 17.9 ± 2.6 mm/year (2004–2008); with a dominant thermosteric impact, the steric sea-level change in the Levantine Sea does not show a clear trend over the same period. However, the thermosteric contribution is also dominant in the Levantine Sea. Ref [72] estimated a SLR of 1.5 mm/yr. in the analysis of tide gauge records from around the North Sea, with modest but not significant differences along its different places. A SLR rate of 3.6 ± 1.6 mm/yr. was calculated in Malacca Strait throughout the period 1986–2013 [73]. The coastlines of the UK exhibited a relative SLR rate of 1–2 mm/yr. as mentioned in [74]. Off the Egyptian Mediterranean coast, the SLR rate has ranged between 2.0 and 3.0 mm/yr. over different periods [75–81].

Altimeter and satellite measurements have proven to be good tools to calculate the relative SLR rate. For the period 1992–2000, altimetry measurements suggested a rapid rising of sea level (20 mm/yr) in the Eastern Mediterranean which has been associated with increases in the sea surface temperature [82]. In [83] a 16-year altimetry data set (1992–2008) was used to investigate the sea-level variations in the Mediterranean Sea. The results revealed that the amplitudes of the annual cycle vary from 4 to 11 cm, except for a small area of value around 16 cm at the southeast of Crete corresponding to the Ierapetra gyre activity. With a combined analysis of altimetry and tide gauge data in the interval 1993–2008, the absolute sea-level rise and crustal motion in the Adriatic Sea were investigated [84]. In the North-Eastern Adriatic, most of the measurements indicated land subsidence with a rate ranging between −0.51 mm/yr. and −0.29 mm/yr. The absolute SLR was 1.9 ± 0.3 mm/yr. in the interval 1993–2008. From 1993 to 2013, the Strait of Malacca exhibited a SLR rate of 4.1 ± 1.9 mm/yr. as revealed by altimeter analysis [85]. Malaysian sea levels have been rising at a spatially variable rate ranging from 1.4 to 4.1 mm/year throughout the period 1993–2008 [86]. Satellite altimetry data were used to assess trends in sea-level rise in the Dumai Sea of Malaysia over 21 years (from 1993 to 2014). The results of the analysis revealed that the SLR rates ranged from 4.80 mm/year to 5.61 mm/year [83]. According to a preliminary examination of SLR rate near Venice, the detected trend by altimetry (4.25 mm/year) is less pronounced than the trends reported by measurements made offshore (5.65 mm/year) and in the lagoon (5.29 mm/year) [26]. Along the western African region, the SLR rate was 2.15 mm/yr. throughout the period 2002–2018 [36]. The regional trends of SLR throughout the period 1993–2019 (**Figure 7**) were depicted in [87].

Impacts from the relative SLR can be assessed and observed in many phenomena, such as storm surges, inundation of seawater, and flooding. Alexandria of Egypt is subsiding at 2 mm/yr. and even without climate change is highly vulnerable to

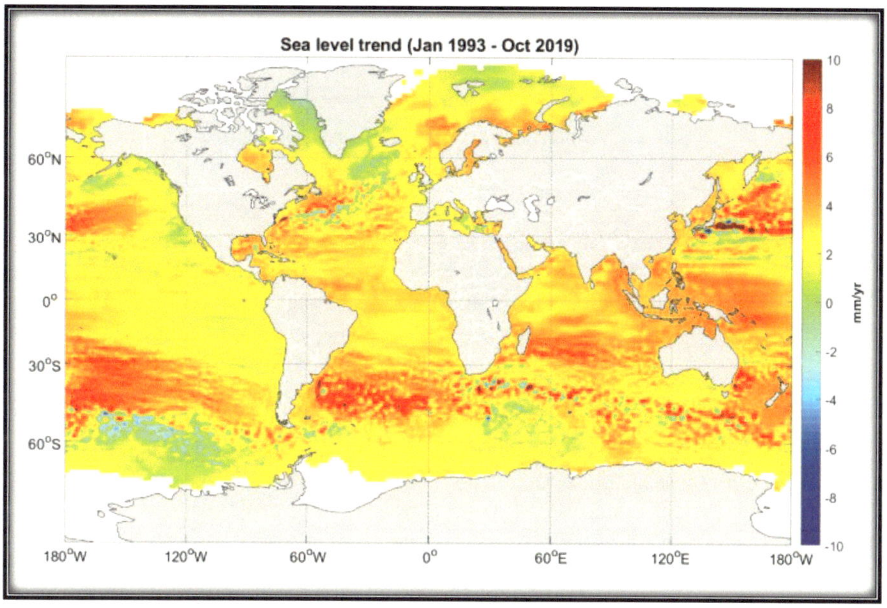

Figure 7.
Regional trends in sea level over 1993 to 2019 from satellite altimetry [87].

flooding and erosion, as 35% (700 km^2) of the land area is below mean sea level [88]. A 50-cm rise in sea level could result in a loss of 13% (0.05 km^2) industrial, 8% (0.46 km^2) urban area, and 1.6% (21 km^2) beach area, and other physical and socioeconomic losses in Port Said governorate (Egypt), costing more than US$2.2 billion [89]. The Nile Delta, Alexandria, Port Said, and Gamassa may suffer not only direct inundation from the SLR but also saltwater intrusion [90]. This will have a direct impact on groundwater resources, soil salinity, agricultural productivity, and quality in the coastal zone. The relative impact of mean SLR in Australia and wind speed in Ireland were examined [91, 92], concluding that SLR has a larger potential than meteorological changes to increase extreme sea levels and flooding probabilities. The higher sea level resulted in increased flooding frequency in several coastal communities, e.g., Boston, Norfolk, and Miami Beach [93, 94]. These frequent flood events, often termed "nuisance flooding," do not cause major damage but do cause material harm, inconvenience, and economic drag. Recently, research [94] used tide gauge data to calculate accumulated flooding time in 12 locations along the Atlantic coast and showed a significant increase in flooding duration. It is suggested that flood duration is a reliable indicator for the accelerating rate of sea-level rise, which is often difficult to estimate on a regional scale. The trends of sea-level extremes due to atmospheric conditions for a period of 150 years (1951–2100), in the Greek seas, under a future climate scenario with highly increasing concentrations of atmospheric greenhouse gases were explored [95]. The results confirmed that the majority of extreme events may appear primarily in winter and secondarily in spring. However, results showed that there is an increase in summer extremes, especially over southern areas due to the increase in cyclogenesis. The damage in the socioeconomical sector in coastal cities affected by the SLR issue typically increases faster than the sea-level rise itself [96]. A vulnerability map of the Egyptian Mediterranean coast to SLR (**Figure 8**) was produced in research [97]. The varying

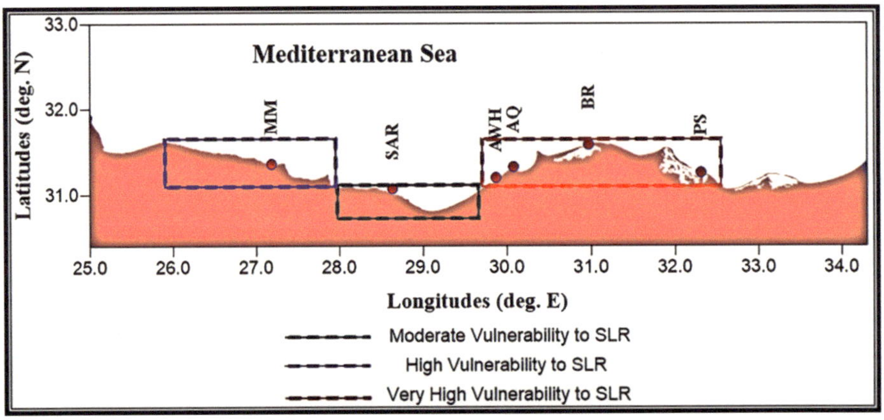

Figure 8.
Vulnerability map of the Egyptian Mediterranean coast to SLR [97].

vulnerability classification to SLR along the Egyptian coast is mainly attributed to the different composition of strata over this coast and the different behavior and rates of land subsidence.

5. Conclusion

To conclude, the sea level has been rising globally. The sea-level rise (SLR) is mainly attributed to two reasons: (1) thermal expansion of seawater and (2) ice melting in the large ice masses on Earth. Both are direct consequences of climate change, which is mainly caused by anthropogenic activities since the industrial epoch. Research and studies have proven that the rate of increase is not uniform all over the globe and that the relative (regional/local) SLR is more important, to discuss the impact of the SLR on coastal structures, the environment, the economy, and human activities. Relative sea-level variations are studied using data from tide gauges and satellites, which are complementary tools to assess these variations. Any shortfall in the sea-level data recorded by gauges can be compensated by that collected from altimetry.

Analyses of tide gauge records indicate that a global mean SLR was between 1.6 and 1.8 mm/yr. over the 20th century. This increased to 3.7 mm/yr. throughout the period 2006–2018. High-precision satellite altimetry suggests a recent global acceleration with a rate of 3.4 mm/yr. Though, this rate varies according to geographical location and land vertical movement around coastlines. Extreme surges, flooding, and seawater intrusion are expected phenomena to associate with the SLR, especially in low-lying coastlines.

The SLR issue has been a main topic of interest in all reports of the Intergovernmental Panel on Climate Change (IPCC) since its First Assessment Report in 1990. According to these reports, there are no solutions to control the SLR but to control climate change through effective adaptation and mitigation plans.

Given that nature has changed over the years and decades and is unlikely to return to its prehuman state anytime soon, the necessity of international cooperation, public awareness campaigns, better monitoring tools, numerical models for simulation and predictions, and the expansion of satellite technology development for marine sciences are all emphasized as key future perspectives.

Author details

Tarek M. El-Geziry
National Institute of Oceanography and Fisheries (NIOF), Egypt

*Address all correspondence to: tarekelgeziry@yahoo.com

References

[1] Pugh DT. Tides, Surges and Mean Sea Level. UK: John Weilly and Sons Ltd.; 1987. p. 472

[2] Davidson M, Edwing R, Blackwell J, Lowell J. Incorporating Sea Level Change Scenarios at the Local Level. USA: NOAA Coastal Services Centre; 2012. p. 14

[3] Jorda G, Gomis D, Alvarez-Fanjul E, Somot S. Atmospheric contribution to Mediterranean and nearby Atlantic sea level variability under different climate change scenarios. Global and Planetary Change. 2012;**80**(81):198-214

[4] IPCC (Intergovernmental Panel on Climate Change). Climate Change 2007: The Physical Science Basis. UK: Cambridge University Press; 2007. p. 996

[5] Pugh D, Woodworth P. Sea-Level Science: Understanding Tides, Surges, Tsunamis and Mean Sea Level. UK: Cambridge University Press; 2014. p. 395

[6] Hicks SD, Sillcox RL, Nichols CR, Via B, McCray EC. Tide and Current Glossary. National Ocean service: US Department of Commerce, NOAA, USA; 2000. p. 29p

[7] El-Geziry TM. Environmental Impact Assessment and Process Simulation of the Tidal Current Energy Resources in the Strait of Messina. Edinburgh: University of Edinburgh; 2010

[8] Bryden IG, Couch SJ, Melville GT. Tidal current resource assessment. Proceedings of the Institution of Mechanical Engineers, Part A: Journal of Power and Energy. 2007;**221**:125-135

[9] Militello A, Zundel AK. Surface water modelling system: Tidal constituents toolbox for ADCIRC. Coastal Engineering Technical Notes. 1999;**4**(21):1-8

[10] Doodson AT. The harmonic development of the tide generation potential. Proceedings of the Royal Society A. 1922;**100**:305-329

[11] Couch SJ. Numerical Modelling of Tidal Flows Around Headlands and Islands [PhD Thesis]. Glasgow: University of Strathclyde; 2001

[12] Pawlowicz R, Beardsley B, Lentz S. Classical tidal harmonic analysis including error estimates in MATLAB using T_TIDE. Computers and Geosciences. 2002;**28**:929-937

[13] Boon J. Secrets of the Tide: Tide and Tidal Current Analysis and Predictions, Storm Surges and Sea Level Trends. UK: Horwood Publishing; 2004. p. 210

[14] IPCC (Intergovernmental Panel on Climate Change). Climate Change 2012: Managing the Risks of Extreme Events and Disasters to Advance Climate Change Adaptation. UK: Cambridge University Press; 2012. p. 582

[15] Palm R, Bolsen T. The science of climate change and sea level. In: Climate Change and Sea Level Rise in South California: The View of Coastal Residents. Cham, Switzerland: Springer; 2020. pp. 5-13

[16] Nerem R, Beckley B, Fasullo J, Hamlington B, Masters D, Mitchum G. Climate-change-driven accelerated sea-level rise detected in the altimeter era. Proceedings of the National Academy of Sciences (PNAS). 2018;**115**(9):2022-2025

[17] IPCC (Intergovernmental Panel on Climate Change). Climate Change

2013: The Physical Science Basis. UK: Cambridge University Press; 2013. p. 1535

[18] IPCC (Intergovernmental Panel on Climate Change). Climate Change 2021: The Physical Science Basis (Summary for Policy Makers). UK: Cambridge University Press; 2021. p. 31p

[19] Archer D. The Climate Crisis: An Introductory Guide to Climate Change. UK: Cambridge University Press; 2010. p. 260p

[20] Fletcher C. Climate Change: What the Science Tells us. USA: Wiley; 2013. p. 265p

[21] Maslin M. Climate Change: A Very Short Introduction. UK: Oxford University Press; 2014. p. 200p

[22] Khedr AM. Sea Level Modeling and Realization of Lowest Astronomical Tide (LAT) for Alexandria Harbor, Egypt [PhD Thesis]. Technology and Maritime Transportation: Arab Academy for Science; 2019

[23] Noye BJ. Tide-well systems I: Some non-linear effects of the conventional tide well. Journal of Marine Research. 1974;**32**(2):129-135

[24] UNESCO. Sustainable Development in Coastal Regions and Small Islands (CSI). Paris: UNESCO Headquarters; 2006. p. 42

[25] El-Geziry TM. Sea-level, tides and residuals in Alexandria Eastern Harbour, Egypt. Egyptian Journal of Aquatic Research. 2021;**47**(1):29-35

[26] Vignudelli S, Birol F, Benveniste J, Fu L-L, Picot N, Raynal M, et al. Satellite altimetry measurements of sea level in the coastal zone. Surveys in Geophysics. 2019;**40**:1319-1349

[27] Nehama FPJ, Veriua ZH, Maueua C, Hibbert A, Calafat F, Cotton PD. Validating sea-level altimetry data against tide gauge for coastal risk analysis in Mozambique. Journal of Marine Science and Engineering. 2022;**10**:1597

[28] Nerem RS, Mitchum GT. Observation of sea level change from satellite altimetry. In: Douglas BC, Kearney MS, Leatherman SP, editors. Sea Level Rise: History and Consequences. California, USA: Elsevier; 2001. pp. 121-163

[29] Cazenave A, Gouzenes Y, Biroll F, Leger F, Passaro M, Calafat F, Shaw A, Nino F, Legeais JF, Oelsmann J, Restano M, Benveniste J. Sea level along the world's coastlines can be measured by a network of virtual altimetry stations. Communications Earth and Environment. 2022;**3**:117

[30] Vignudelli S, Vignudelli S, Kostianoy AG, Cipollini P, Benveniste J. Coastal Altimetry. Berlin: Springer; 2011. p. 362p

[31] Fu L, Cazenave A. Satellite Altimetry and Earth Sciences: A Handbook for Techniques and Applications. Academic Press; 2000. p. 463

[32] Cazenave A, Llovel W. Contemporary sea level rise. Annual Review of Marine Science. 2010;**2010**(2):145-173

[33] Nicholls RJ, Cazenave A. Sea level change and the impacts in coastal zones. Science. 2010;**328**:1517-1520

[34] Church JA, White NJ. Sea-level rise from the late 19th to the early 21st century. Surveys in Geophysics. 2011;**4**(5):585-602

[35] Hay CC, Morrow E, Kopp RE, Mitrovica JX. Probabilistic reanalysis of twentieth-century sea-level rise. Nature. 2015;**517**:481-484

[36] Benveniste J, Birol F, Calafat F, Cazenave A, Dieng H, Gouzenes Y, et al.

Coastal sea level anomalies and associated trends from Jason satellite altimetry over 2002-2018. Nature Scientific Data. 2020;7:357

[37] Meyssignac B, Cazenave A. Sea level: A review of present-day and recent-past changes and variability. Journal of Geodynamics. 2012;**58**:96-109

[38] de Lange P, Carter R. Sea-level Change: Living with Uncertainty. The Global Warming Policy Foundation; 2014 42p

[39] Hansen J, Sato M, Hearty P, Ruedy R, Kelley M, Masson-Delmotte V, et al. Ice melt, sea level rise and super storms: Evidence from paleoclimate data, climate modeling, and modern observations that 2°C global warming could be dangerous. Atmospheric Chemistry and Physics. 2016;**16**:3761-3812

[40] Mengel M, Levermann A, Frieler K, Robinson A, Marzeion B, Winkelmann R. Future sea-level rise constrained by observations and long-term commitment. Proceedings of the National Academy of Sciences (PNAS). 2016;**113**(10):2597-2602

[41] Allison I, Alley RB, Fricker HA, Thomas RH, Warner RC. Ice sheet mass balance and sea level. Antarctic Science. 2009;**21**:413-426

[42] Rignot E, Box JE, Burgess E, Hanna E. Mass balance of the Greenland ice sheet from 1958 to 2007. Geophysical Research Letters. 2008;**35**:L20502

[43] Velicogna I. Increasing rates of ice mass loss from the Greenland and Antarctic ice sheets revealed by GRACE. Geophysical Research Letters. 2009;**36**:L2009

[44] Church JA, Woodworth PL, Aarup T, Wilson WS. Understanding Sea Level Rise and Variability. Wiley-Blackwell: UK; 2010. p. 456p

[45] Marcos M, Tsimplis MN, Shaw AGP. Sea level extremes in southern Europe. Journal of Geophysical Research. 2009;**114**:C01007

[46] Cid A, Menéndez M, Castanedo S, Abascal AJ, Méndez FJ, Medina R. Long-term changes in the frequency, intensity and duration of extreme storm surge events in southern Europe. Climate Dynamics. 2016;**46**(5):1502-1516

[47] Lionello P, Myfato R, Tomasin A. Sensitivity of free and forced oscillations of the Adriatic Sea to sea level rise. Climate Research. 2005;**29**:23-39

[48] Obeysekera J, Park J. Scenario-based projections of extreme sea levels. Journal of Coastal Research. 2013;**29**(1):1-7

[49] Lowe JA, Gregory JM, Flather RA. Changes in the occurrence of storm surges around the United Kingdom under a future climate scenario using a dynamic storm surge model driven by the Hadley Centre climate models. Climate Dynamics. 2001;**18**:179-188

[50] Demernard J, Saetra O, Roed LP. Future wind, wave and storm surge climate in the northern North Atlantic. Climate Research. 2002;**23**:39-49

[51] Danard M, Munro A, Murty T. Storm surge hazard in Canada. Natural Hazards. 2003;**28**:407-431

[52] Gonnert G. Maximum storm surge curve due to global warming for the European North Sea region during the 20th - 21st century. Natural Hazards. 2004;**32**:211-218

[53] Parris A, Bromirski P, Burkett V, Cayan D, Culver M, Hall J, et al. Global Sea Level Rise Scenarios for the US

National Climate Assessment. NOAA Technical Memo, OAR CPO-1. 2012. p. 29

[54] IPCC (Intergovernmental Panel on Climate Change): Climate Change 2014. Impacts, Adaptation, and Vulnerability. UK: Cambridge University Press; 2014. p. 1132

[55] Kopp RE, Horton RM, Little CM, Mitrovica JX, Oppenheimer M, Rasmussen DJ, et al. Probabilistic 21st and 22nd century sea-level projections at a global network of tide-gauge sites. Earth's Future. 2014;**2**:383-406

[56] El-Sharkawy H, Rashed H, Rached I. Climate changes: The impacts of sea level rise on Egypt. In: Proceedings of the 45th ISOCARP Congress. 2009. p. 11

[57] Bindoff NL, Willebrand J, Artale V, Cazenave A, Gregory J, Gulev S, et al. Observations: Oceanic climate change and sea level. In: Climate Change 2007: The Physical Science Basis. Contribution of WG I to IPCC AR4. Cambridge, UK: Cambridge University Press; 2007. pp. 385-432

[58] Wunsch C, Ponte RM, Heimbach P. Decadal trends in sea level patterns: 1993-2004. Journal of Climate. 2007;**20**(24):5889-5911

[59] Lombard A, Garric G, Penduff T, Molines JM. Regional variability of sea level change using a global ocean model at ¼° resolution. Ocean Dynamics. 2009;**3**:433-449

[60] Levitus S, Antonov JL, Boyer TP, Locarnini RA, Garcia HE, Mishonov AV. Global Ocean heat content 1955-2008 in light of recently revealed instrumentation. Geophysical Research Letters. 2009;**36**:L07608

[61] Stammer D. Response of the global ocean to Greenland and Antarctica melting. Journal Geophysical Research. 2008;**113**:C06022

[62] Stammer D, Agarwal N, Herrmann P, Köhl A, Mechoso CR. Response of a coupled ocean–atmosphere model to Greenland ice melting. Surveys in Geophysics. 2011;**32**:621-642

[63] Okumura YM, Deser C, Hu A, Timmermann A, Xie SP. North Pacific climate response to freshwater forcing in the Subarctic North Atlantic: Oceanic and atmospheric pathways. Journal of Climate. 2009;**22**:1424-1445

[64] Tsimplis MN, Baker TF. Sea level drop in the Mediterranean Sea: An indicator of deep water salinity and temperature changes. Geophysical Research Letters. 2000;**27**(12):1731-1734

[65] Calvo E, Simó R, Coma R, Ribes M, Pascual J, Sabatés A, et al. Effects of climate change on Mediterranean marine ecosystems. Climate Research. 2011;**50**:1-29

[66] Bonaduce A, Pinardi N, Oddo P, Spada G, Larnicol G. Sea-level variability in the Mediterranean Sea from altimetry and tide gauges Climate Dynamics. 2016;**47**:2851-2866

[67] Shirman B. Israel coast sea level changes during 1958-2001. Survey of Israel. 2001;**Report No. RD4/1**:17p

[68] Rosen DS. Long term remedial measures of sedimentological impact due to coastal developments on the southeastern Mediterranean coast. In: Proceedings of Littoral 2002, The Changing Coast (EUROCOAST/EUCC, Ed), September 2002. Porto, Portugal; 2002. pp. 322-331

[69] Klein M, Lichter M, Zviely D. Recent sea level changes along Israeli

and Mediterranean coasts. Horizons in Geography. 2004;**60**(61):167-176

[70] Tsimplis M, Marcos M, Somot S. 21st century Mediterranean sea level rise: Steric and atmospheric pressure contributions from a regional model. Global and Planetary Change. 2008;**63**:105-111

[71] Passaro M, Seitz F. Steric sea level variations in the central-eastern Mediterranean Sea from Argo observations. Bollettino di Geofisica Teorica ed Applicata. 2012;**52**(1):131-147

[72] Wahl T, Haigh ID, Woodworth PL, Albrecht F, Dillingh D, Jensen J, et al. Observed mean sea level changes around the North Sea coastline from 1800 to present. Earth-Science Reviews. 2013;**124**:51-67

[73] Luu QH, Tkalich P, Tay TW. Sea level trend and variability around Peninsular Malaysia. Ocean Science. 2015;**11**(4):617-628

[74] Horsburgh K, Rennie A, Palmer M. Impacts of climate change on sea-level rise relevant to the coastal and marine environment around the UK. MCCIP Science Review. 2020;**2020**:116-131

[75] Frihy OE. Sea-level rise and shoreline retreat of the Nile delta promontories, Egypt. Natural Hazards. 1992;**5**:65-81

[76] Frihy OE. The Nile delta-Alexandria coast: Vulnerability to Sea-level rise, consequences and adaptation. Mitigation and Adaptation Strategies for Global Change. 2003;**8**:115-138

[77] Shaker AA, Saad AA, Al-Naggar D, Faisal H. Absolute Sea-level Rise Estimation at Alexandria using Tide Records and GPS Observations. Vol. 2011. Marrakech, Morocco: FIG Working Week 2011; 2011. p. 14

[78] Maiyza IA, El-Geziry TM. Long term sea-level variation in the south-eastern Mediterranean Sea: A new approach of examination. Journal of Operational Oceanography. 2012;**5**(2):53-59

[79] Hendy DM. Risk Reduction of Extreme Climate Changes on the Coastal Zone of Alexandria, Egypt [PhD Thesis]. Institute of Graduate Studies and Research (IGSR): Alexandria University; 2018

[80] El-Geziry TM, Said MA. Spatial variations of sea level along the egyptian mediterranean coast. Athens Journal of Mediterranean Studies. 2020;**6**(2):141-154

[81] El-Geziry TM, El-Wakeel YM. A decadal sea-level variability in Port-Said Harbour (Egypt). Egyptian Journal of Aquatic Research. 2023;**49**(1):33-39

[82] Cazenave A, Cabanes C, Dominh K, Mangiarotti S. Recent sea level changes in the Mediterranean Sea revealed by TOPEX/POSEIDON satellite altimetry. Geophysical Research Letters. 2001;**28**(8):1607-1610

[83] Vigo MI, Sánchez-Realesa JM, Trottini M, Chaoc BF. Mediterranean Sea level variations: Analysis of the satellite altimetric data, 1992-2008. Journal of Geodynamics. 2011;**52**:271-278

[84] Fenoglio-Marc L, Braitenberg C, Tunini L. Sea level variability and trends in the Adriatic Sea in 1993-2008 from tide gauges and satellite altimetry. Physics and Chemistry of the Earth. 2012;**40**(41):47-58

[85] Ariana D, Permana C, Setiawan Y. Study of sea level rise using satellite altimetry data in the sea of Dumai, Riau. Indonesia. Journal of Geomatics and Planning. 2017;**4**(1):75-82

[86] Din AHM, Omar KM, Naeije M, Ses S. Long-term sea level change in

the Malaysian seas from multi-mission altimetry data. International Journal of Physical Sciences. 2012;7(10):1694-1712

[87] Cazenave A. Climate change and sea level rise. Spatium. 2020;**46**:1-15

[88] El-Raey M. Vulnerability assessment of the coastal zone of the Nile delta, Egypt, to the impacts of sea level rise. Ocean and Coastal Management. 1997;**37**(1):29-40

[89] Agrawala S, Moehner A, El-Raey M, Conway D, van Aalst M, Hagenstad M, et al. Development and climate change in Egypt: Focus on coastal resources and the nile. Organisation for Economic Co-operation and Development (OECD). COM/ENV/EPOC/DCD/DAC(2004)1/FINAL. 2004. p. 68

[90] El-Raey M. Impact of Sea Level Rise on the Arab Region. 2009. Available from: https://research.fit.edu/media/site-specific/researchfitedu/coast-climate-adaptation-library/africa/morocco-algeria-tunisia/El-Raey.-2010.-SLR-Impacts-on-Arab-Region.pdf [Accessed: April 19, 2023]

[91] McInnes KL, Macadam I, O'Grady JG. The Effect of Climate Change on Extreme Sea Levels Along Victoria's Coast. Australia: CSIRO; 2009. p. 55p

[92] Brown S, Kebede AS, Nicholls RJ. Sea-Level Rise and Impacts in Africa, 2000 to 2100. Southampton, UK: School of Civil Engineering and the Environment, University of Southampton; 2011. p. 215. Available from: https://www.researchgate.net/publication/307210112_Sea-Level_Rise_and_Impacts_in_Africa_2000_to_2100 [Accessed: May 27, 2023]

[93] Ezer T, Atkinson LP, Corlett WB, Blanco JL. Gulf Stream's induced sea level rise and variability along the U.S. mid-Atlantic coast. Journal of Geophysical Research. 2013;**118**:685-697

[94] Ezer T, Atkinson LP. Accelerated flooding along the U.S. East Coast: On the impact of sea level rise, tides, storms, the Gulf Stream, and the North Atlantic Oscillations. Earth's Futures. 2014;**2**:362-382

[95] Makris C, Androulidakis Y, Krestinitis Y, Kombiadou K, Baltikas V. Numerical modelling of storm surges in the Mediterranean Sea under climate change. In: Proceedings of the 36th IAHR World Congress, 28 June-3 July 2015. the Netherlands; 2015. pp. 1-13

[96] Boettle M, Rybski D, Kropp JP. Quantifying the effect of sea level rise and flood defence: A point process perspective on coastal flood damage. Natural Hazards and Earth System Science. 2016;**16**:559-576

[97] El-Geziry TM. On the vulnerability of the Egyptian mediterranean coast to the sea level rise. Athens Journal of Science. 2020;**7**(4):195-206

Chapter 4

Bathymetry Estimation from Satellite Altimeter-Derived Gravity Data

Ljerka Vrdoljak and Tomislav Bašić

Abstract

Bathymetry underpins all marine and ocean research. It is common knowledge that there is a global deficit of high-resolution bathymetry based on modern acoustic techniques. Satellite altimetry enabled modeling of the global seafloor topography and revealed new morphological features in the unmapped areas of the oceans and seas. This chapter gives an overview of the physical problem and different approaches to estimating the bathymetry from satellite altimeter-derived gravity data. Characteristics of recent versions of frequently used global bathymetry models are presented. Moreover, this chapter demonstrates the possibility of regional bathymetry modeling by the gravity-geologic method in the Adriatic Sea.

Keywords: bathymetry mapping, global bathymetry grids, gravity anomalies, gravity-geologic method, regional bathymetry modeling

1. Introduction

Bathymetry is an important input parameter or a frame that supports all marine research. Although there are global and regional initiatives to improve our understanding of seafloor topography [1–3], less than 25% of the world's seas have been mapped with high resolution that is able to identify features of a few tens of meters in size [1]. Current global seafloor topography is estimated from altimeter data and augmented with available grids from a variety of techniques, mainly shipborne depth soundings [1, 2, 4–7]. As compared to modern acoustic techniques, bathymetry derived from altimetry has a coarse spatial resolution [8]. However, the data from altimeter missions enabled revelling of buried and unmapped features of global seafloor topography [4, 9]. Altimeter data supplemented the sparse shipborne soundings and improved our knowledge of the seafloor topography by bathymetry inversion from altimeter-derived gravity anomalies. Global marine gravity grids formed from high-density altimeter data (e.g. [10, 11]) and digital data bases of shipborne soundings (e.g. [10]) enabled estimation of global seafloor topography [11].

This chapter gives an overview of the relationship between the topography of the seafloor and gravity. Diverse approaches to estimate the bathymetry from altimeter-derived gravity data, in space and frequency domain, are briefly presented.

IntechOpen

Characteristics of frequently used global bathymetry models are depicted. Moreover, the chapter demonstrated the possibility of regional bathymetry modeling by the gravity-geologic method (GGM) in the Adriatic Sea. The estimated bathymetry grid was compared to global grids in the study area, and their quality was assessed as compared to chart soundings.

2. The relationship between depth and gravity

The depth variations of the seafloor can be observed as height variations of mass elements of the density $\Delta\rho$ which is the contrast between the density of the seafloor ρ_c and seawater ρ_w [12]. The result of the seafloor topography variation is the disturbance in the local gravity field.

The disturbing potential T(r) due to mass element of the volume V and density $\Delta\rho$ is [12]:

$$T(r) = G\Delta\rho \int_V \frac{dV}{|r - r'|} \tag{1}$$

where G is the gravitational constant, r is the coordinate vector of location, and r′ is the coordinate vector of the center of the mass element.

The geoid undulation N is related to the disturbing potential T by Brun's formula [11, 12]:

$$N \cong \frac{1}{g_0} T \tag{2}$$

where g_0 is the average acceleration of gravity regarding the geodetic latitude.

The gravity anomaly Δg is the vertical derivate of the disturbing potential [11, 12]:

$$\Delta g = -\frac{\partial T}{\partial z} \tag{3}$$

The east and the north component of vertical deflection represent the slope of the geoid in x and y direction:

$$\eta = -\frac{1}{g_0}\frac{\partial T}{\partial x}, \xi = -\frac{1}{g_0}\frac{\partial T}{\partial y} \tag{4}$$

Laplace's equation links these quantities together [11, 12]:

$$\frac{\partial \eta}{\partial x} + \frac{\partial \xi}{\partial y} = -\frac{\partial \Delta g}{\partial z} \tag{5}$$

Disaggregating of the computation area in Eq. (1) into discretized elements of surface $\Delta\Omega(r')$ and regarding the Eq. (2), topography undulation N(r) is given by [12]:

$$N(r) = \frac{G}{g_0}\Delta\rho \sum_{r'} \Delta\Omega(r') \int_{z_b}^{z_t} \frac{dz}{|r - r'|} \tag{6}$$

where z_b i z_t are depth on the bottom and top of the mass element.

In spectral domain, relationship between topography of the seafloor and gravity anomalies is [13]:

$$\mathrm{F}[\Delta g] = 2\pi G(\rho_c - \rho_w)e^{-2\pi kd} \sum_{n=1}^{\infty} \frac{(2\pi k)^{n-1}}{n!} \mathrm{F}[h^n] \tag{7}$$

where F[ ] is the two-dimensional Fourier transform operator, k is the wave number; $k = \sqrt{k_x^2 + k_y^2}$ where $k_x=1/\lambda_x$, a $k_y=1/\lambda_y$, λ_x i λ_y are wavelengths at x and y direction, and h is depth of the seafloor located at the mean sea depth d.

There are several inverse approaches to model topography of the seafloor from altimeter-derived gravity anomalies [12].

In this study, two commonly used approaches are reviewed, Smith and Sandwell (S&S) in frequency domain and gravity-geologic method (GGM) in space domain.

2.1 Smith and Sandwell approach (S&S)

Smith and Sandwell [4, 9, 11] suggested that a correlation between variations in altimeter-derived gravity anomalies and topography of the seafloor can be found in the wavelength band of 15–200 km. If variations in seafloor undulations are much smaller than mean sea depth, Eq. (7) can be limited to the first term [11]:

$$\mathrm{G}(k) = 2\pi G(\Delta\rho)e^{-2\pi kd}\mathrm{H}(k) = \mathrm{Z}(k)\mathrm{H}(k) \tag{8}$$

$$\mathrm{H}(k) = Z^{-1}(k)\mathrm{G}(k) \tag{9}$$

where $\mathrm{G}(k)$ is a Fourier transform of the gravity anomalies, $\mathrm{H}(k)$ is a Fourier transform of the seafloor topography, and $\mathrm{Z}(k)$ is the isotropic transfer or the admittance function.

The main steps in the S&S approach are as follows [4, 9, 11]:

The base bathymetry grid in frequency domain $\mathrm{H_B}(k)$ is separated into low-pass (long-wavelength) bathymetry $\mathrm{H_L}(k)$ and high-pass (short-wavelength) bathymetry $\mathrm{H_S}(k)$ components using a Gaussian filter.

Gravity anomalies in the frequency domain $\mathrm{G}(k)$ are band-pass filtered and downward continued using the Wiener filter $\mathrm{W}(k)$ to stabilize the procedure:

$$\mathrm{G_{BP}}(k) = \mathrm{G}(k)\ \mathrm{W}(k)\ e^{2\pi kd} \tag{10}$$

The Wiener filter is composed of high-pass filter $\mathrm{W_1}(k)$ and low-pass filter $\mathrm{W_2}(k)$ whose original forms are defined by Smith and Sandwell [4].

The band-passed filtered bathymetry $\mathrm{H_{BP}}(k)$ is obtained by applying the filter to base bathymetry grid in the frequency domain.

According to the admittance theory [14], the relationship between gravity and topography is linear, so topography can be inverted from gravity by simply multiplying with theoretical topography/ratio scaling factor $\mathrm{S_T} = (2\pi G\Delta\rho)^{-1}$ [12]. Instead of using the theoretical value, in overlapping area, a robust regression analyse is performed between band-passed bathymetry $\mathrm{H_{BP}}(k)$ and band-passed gravity anomalies $\mathrm{G_{BP}}(k)$ to estimate the topography/ratio scaling S.

The total predicted bathymetry by S&S approach in the space domain $d_p(x)$ is

$$d_P(x) = d_L(x) + S\,g(x) + d_S(x) \tag{11}$$

where $d_L(x)$ and $d_S(x)$ are the spatial domain of the low-passed bathymetry $H_L(k)$ and high-passed bathymetry $H_S(k)$, respectively, and $g(x)$ is a spatial domain of band-passed gravity $G_{BP}(k)$.

2.2 Gravity-geologic method (GGM)

Although the gravity-geologic method (GGM) was originally used to determine the depth of a glacial sediment above the bedrock [15], it has been adopted and utilized in recent studies to estimate the regional bathymetry from altimetry [16–20].

The observed free-air gravity anomalies at the sea surface Δg can be separated to the referent, long-wavelength gravity Δg_{long} caused by the distribution of masses deep inside the Earth's body and the residual, short-wavelength gravity field Δg_{short} caused by the distribution of masses above the datum D. Datum D is usually determined as the deepest depth.

The GGM calculates the residual field from a Bouguer slab formula using the control soundings d_j:

$$\Delta g_{short}(j) = 2\pi G\Delta\rho\left(d_j - D\right) \tag{12}$$

where G is the gravitational constant and $\Delta\rho$ is the density contrast between seafloor and seawater.

The long-wavelength gravity field in the known points $\Delta g_{long}(j)$ is determined by the simple subtraction:

$$\Delta g_{long}(j) = \Delta g(j) - \Delta g_{short}(j) \tag{13}$$

The long-wavelength gravity is then interpolated to the unknown i-th points from the known $\Delta g_{long}(j)$ at known j-th points. The short-wavelength gravity $\Delta g_{short}(i)$ at unknown i-th points is calculated by subtracting the long-wavelength gravity $\Delta g_{long}(i)$ from the observed gravity $\Delta g(i)$:

$$\Delta g_{short}(i) = \Delta g(i) - \Delta g_{long}(i) \tag{14}$$

Depth at the unknown points d_i is determined by simple inversion of the Eq. (12):

$$d_i = \frac{\Delta g_{short}(i)}{2\pi G\Delta\rho} + D \tag{15}$$

3. Global bathymetry models

Global bathymetry models have been constructed based on satellite altimetry, employing different data and techniques. **Table 1** presents a summary of attributes of recognized and frequently used global bathymetry models (recent version): (1) DTU10BAT (Bathymetry model from Space Institute of the Technical University of Denmark) [26], (2) ETOPO 1 (National Oceanic and Atmospheric Administration ETOPO 1 Arc-Minute Global Relief Model) [7], (3) GEBCO 2021 (The General Bathymetric Chart of the Ocean) [6], (4) SRTM 15+ v2.3 (Shuttle Radar Topography

DBM	DTU10BAT	ETOPO 1	GEBCO 2021	SRTM15+ v2.3	SS v20.1
Grid Spacing	1′–2′ (Equator)	1′	15″	15″	1′
Release Year	2010	2009	2021	2021	2020
Based on	Altimeter-derived gravity DTU10 and ship depth soundings	±80° latitude 2 arc min SS grid (2008)	SRTM15+ v2.2 augmented with additional bathymetry	Altimeter-derived gravity and ship depth soundings	Altimeter-derived gravity and ship depth soundings
Website	[21]	[22]	[23]	[24]	[25]

Table 1.
Global bathymetry models relying on satellite altimetry.

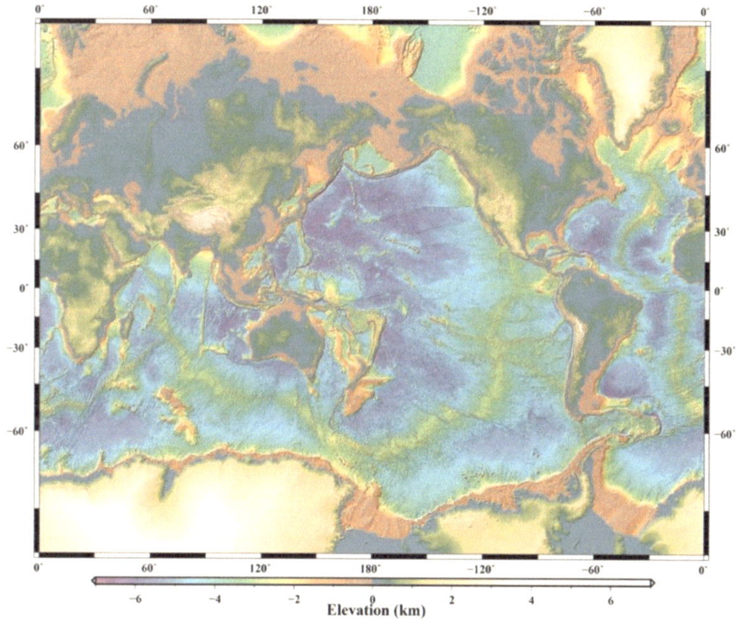

Figure 1.
Global bathymetry and topography at 15 arc seconds [5].

Mission: Global Bathymetry and Topography at 15 arc seconds) [5], and (5) SS v20.1 (Global topography from Scripps Institution of Oceanography) [4].

Global seafloor topography (**Figure 1**) relies on bathymetry estimated from altimeter-derived gravity anomalies, employing the S&S approach in the frequency domain adjusted for digital data processing. The base bathymetry layer is afterwards augmented by bathymetric data from other in situ or remote sensing techniques and existing composite bathymetry grids.

Several studies evaluated and compared available bathymetry grids on a global and regional scale [27–30]. Differences between grids resulted from different density, distribution and accuracy of the input bathymetry, grid misregistration, data smoothing, and integration of different datasets to form the global grid [27, 30]. Quality of depth estimated from the altimeter derived gravity is related to limitations of the altimeter technology, causing robust bathymetry due to the noise in the solution [27, 29] and large discrepancies in coastal areas [30]. SS global bathymetry model provided a base bathymetry layer for most of global and regional bathymetry solutions. SS model reflects state of the art in marine gravity modeling [31]. Combined with a large database of shipborne surveys at Scripps Institution of Oceanography, the SS model is continuously upgraded and generally considered to be a reliable and up-to-date bathymetry source [27]. However, an uneven distribution of sparse in situ bathymetric data can result in large depth anomalies in the inversion of the seafloor topography. On a global scale, depth uncertainty can be expected to be less than 100 meters in deep ocean areas and greater than 100 meters between the shoreline and the continental rise [5, 28].

4. Regional bathymetry modeling: A case study of Adriatic Sea

There is an ongoing effort by the scientific community to improve bathymetry solutions on global and regional scale [1, 2]. Base bathymetry estimated from altimeter-derived gravity is augmented with high-quality survey grids or composite bathymetry products. The GGM method has been successfully utilized for regional bathymetry modeling in different marine regions [16–20, 32, 33]. The difference between the quality of models derived from the GGM and the S&S approach is negligible, as it is more dependent on the availability of the shipborne soundings [33]. The GGM method has an algorithm in the spatial domain, so there is no need for transformation to a frequency domain, but the accuracy of the method depends on the density and distribution of shipborne soundings, and the estimation of a density contrast between the seafloor and seawater [33].

In this study, a 1/16′ by 1/16′ bathymetry model of the Adriatic was constructed by the GGM method. The base model was augmented by the *in situ* soundings from EMODnet network and nautical charts. The model was compared to the global solutions listed in Par. 3, and the quality of the models was estimated regarding chart soundings.

4.1 Study area and datasets

4.1.1 Study area

The Adriatic Sea (12° 3′ – 20° 1′ E, 39° 44′ - 45° 48′ N) is the most northern part of the Mediterranean Sea connected to the Ionian Sea via the Strait of Otranto. Limits of the Adriatic Sea and land mask were adopted from IHO and the Flanders Institute [34, 35]. The Adriatic Sea is a shallow sea with a median depth of 100 meters [36]. By bathymetry, the Adriatic is divided into three sub-basins: the shallowest North sub-basin, the transitional zone of the Middle sub-basin, and the South sub-basin that comprises the South Adriatic Pitt, the deepest part of the Adriatic with depths extending under 1200 meters (**Figure 2**) [36].

4.1.2 Altimeter-derived gravity anomalies

This study explores the possibility of inverting bathymetry from altimeter-derived gravity anomalies by the GGM method in the Adriatic Sea. Models of free-air gravity

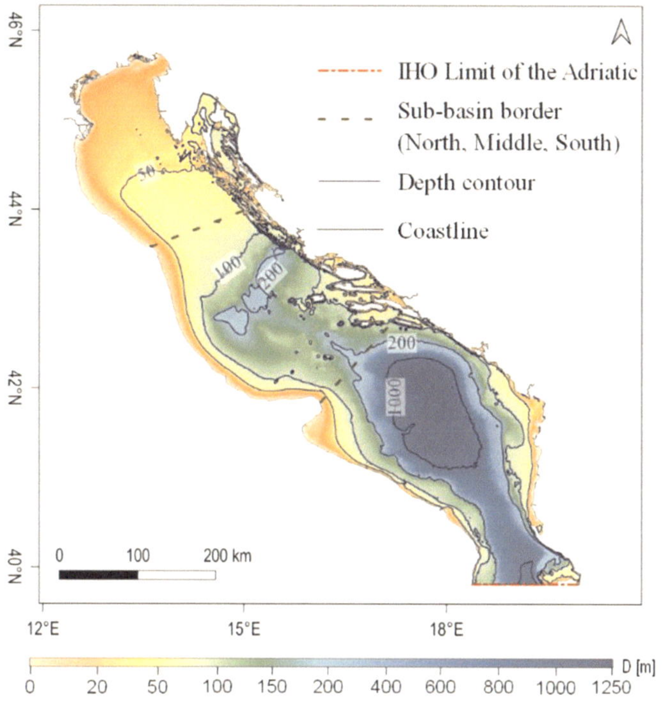

Figure 2.
Adriatic Sea. (Bathymetry source [6]).

Δg	**DTU10** [10^{-5} **ms**$^{-2}$]	**SS v.29.1** [10^{-5} **ms**$^{-2}$]
MIN	−116.22	−135.40
MAX	115.14	129.80
MEAN	−15.20	−15.65
σ	35.52	36.90

Table 2.
Statistic of gravity anomalies in the Adriatic from DTU 10 and SS v 29.1 models: minimum (MIN), maximum (MAX), mean and standard deviation (σ).

anomalies from the Technical University of Denmark, DTU10 model [26], and from Scripps Institution of Oceanography, SS v. 29.1 [31] were used. General statistics of models in the study area is presented in **Table 2**.

The current accuracy of gravity anomalies derived from altimeter data is around 2×10^{-5} ms^{-2} [31]. As presented in **Figure 3c**, the largest differences between models (>40×10^{-5} ms^{-2}) were along well-indented eastern Adriatic coast.

4.1.3 Control and check soundings

Control bathymetry was composed from EMODnet 2020 bathymetry [2] in the western Adriatic, GEBCO One Minute Grid [37] in the south-eastern Adriatic and soundings from nautical charts in the eastern Adriatic (**Figure 4a**). Control bathymetry,

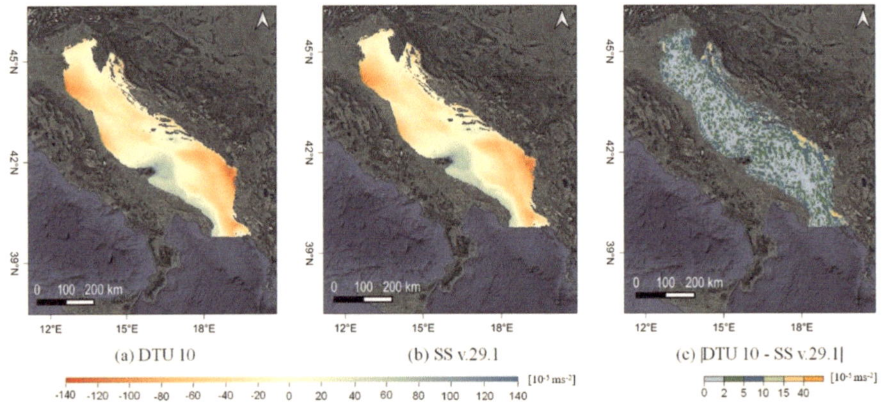

Figure 3.
Free-air gravity anomalies over Adriatic Sea: (a) DTU10 gravity anomaly, (b) SS v29.1 gravity anomaly, and (c) absolute difference between models [26, 31].

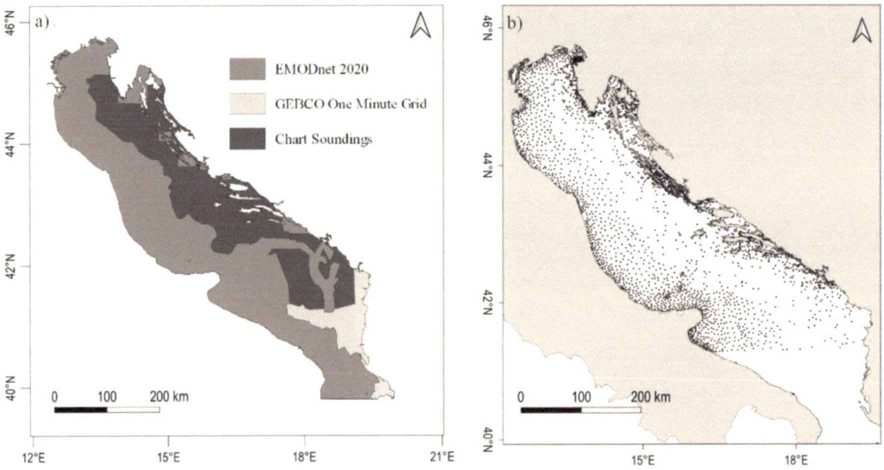

Figure 4.
(a) Control soundings and (b) check soundings.

needed for accurate modeling of the referent gravity field, consisted of 45 666 depths. Over 3500 soundings from nautical charts were used as check soundings needed for the estimation of a density contrast and quality control (**Figure 4b**). The quality of control and check soundings was estimated to be better than 2 + 0.05% depth meter [2].

4.2 Methodology

4.2.1 Bathymetric recipe

Bathymetry was calculated in three steps:

1. The first step was constructing the base bathymetric layer. The 1 arc-minute base bathymetric grid was estimated from gravity anomalies using the gravity-geologic method (GGM) (Par. 2. 2).

2. Differences between the control soundings and base bathymetry were derived. Gaps between points, at a distance larger than 1 arc minute from a point, were filled with zero values to prevent the generation of artificial morphology [5]. Differences were gridded to a model with a 1/16 arc-minute grid spacing.

3. The base bathymetry layer was re-interpolated to 1/16 arc-minute grid spacing using a bilinear interpolation. The final bathymetry model resulted in adding the differences to the re-interpolated base bathymetry.

4.2.2 Comparison and quality assessment of bathymetric grid

For pixel-to-pixel comparison between the calculated model of the Adriatic Sea and available global grids in the study area, global grids were resampled to a grid spacing of 1/16 arc-minute by bilinear interpolation. Absolute differences between the calculated digital bathymetry model (DBM) and global models in identical points were calculated and analyzed.

Residuals between check soundings and model were taken as a measure of model accuracy. With the most widely used measure for quality assessment root mean square error (RMSE), a normalized root mean square error (NRMSE) was calculated in different depth ranges.

4.3 Result

4.3.1 Digital bathymetry model of the Adriatic Sea GGM+ DBM

Bathymetry of the Adriatic Sea was estimated from altimeter-derived gravity anomalies by the GGM method, using a theoretical density contrast between the seafloor topography and the seawater of 1670 kgm^{-3}. The bathymetric model inverted from DTU 10 gravity anomalies has the RMSE of 25.41 m, while bathymetry estimated from SS v 29.1 gravity anomalies has the RMSE of 30.05 m. The tuning density contrast, which minimized the RMSE of the predicted bathymetry, was estimated by a trade-off diagram. As shown in **Figure 5**, a density contrast higher than 10^4 kgm^{-3} stabilized the trade-off diagram around RMSE of 14 m and a correlation (r_P) of 99.60%.

Tuning density contrast of 15 000 kgm^{-3} was chosen to predict bathymetry by the GGM method in the Adriatic Sea. The digital bathymetry model (DBM) derived from DTU10 gravity anomalies (DTU10 DBM_{15}) had the RMSE of 13.80 m. The RMSE of a

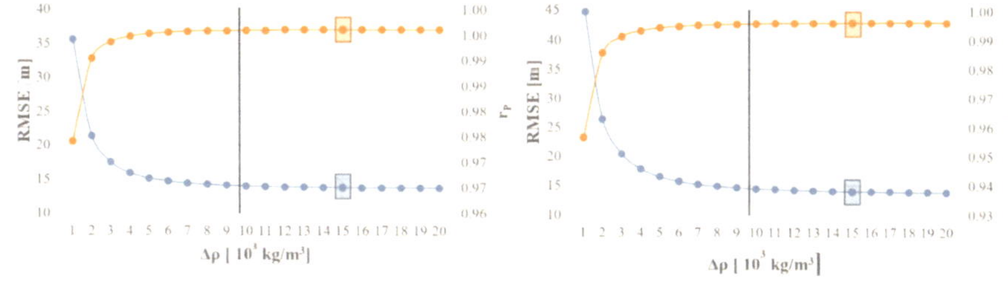

Figure 5.
Trade-off diagram for selecting a tuning density contrast in the study area for bathymetry modeling by GGM method (a) from DTU 10 gravity anomalies and (b) from SS v.29.1 gravity anomalies.

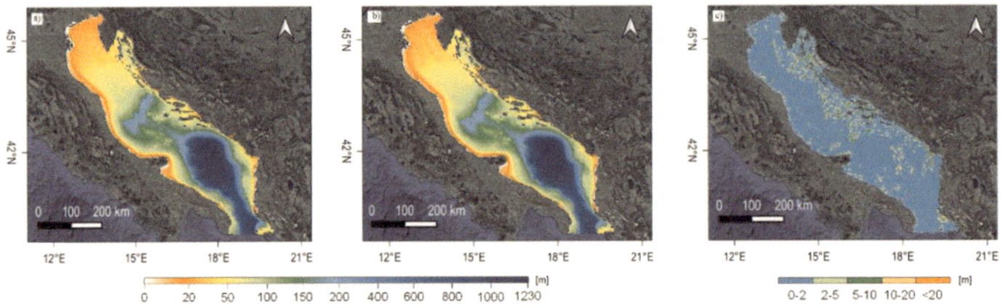

Figure 6.
Digital bathymetry models estimated from altimeter-derived gravity anomalies: (a) DTU10 DBM_{15}, (b) SS DBM_{15}, and (c) absolute differences between models.

	Depth[m]			
DBM	**Max**	**Mean**	**Median**	**σ**
DTU 10 DBM_{15}	1225	250	100	342
SS DBM_{15}	1224	250	100	342

Table 3.
Summary statistics of DTU10 DBM_{15} and SS DBM_{15} digital bathymetric models (DBM): maximum depth (Max), mean depth (Mean), median depth (Median), and standard deviation od depth (σ).

DBM derived from SS v 29.1 gravity anomalies (SS DBM_{15}) was 14.0 m. **Figure 6** represents 1 arc-minute DBMs of the Adriatic referring to the Mean Sea Level (MSL). DBMs were derived from DTU10 and SS v 29.1 gravity anomalies by the GGM method using the tuning density contrast of 15 000 kgm^{-3}.

Summary statistics of DTU10 DBM_{15} and SS DBM_{15} models over study area is presented in **Table 3**.

DTU10 DBM_{15} and SS DBM_{15} had a high degree of correspondence due to the equal tuning density contrast for bathymetry inversion by the GGM method. The largest discrepancies were in the coastal area along the eastern coast (**Figure 6c**). That resulted from the differences between input gravity models.

DTU10 DBM_{15} bathymetric model has a slightly lower RMSE compared to SS $DBM_{15.}$ Therefore, it was chosen as a base bathymetric layer to compute an enhanced bathymetry model of the Adriatic Sea (GGM+ DBM). If possible, pixel values were reset to the value of directly observed bathymetry. A modified Remove-Restore procedure was applied [5]. GGM+ DBM with 1/16 arc-minute grid spacing was enhanced by the EMODnet 2020 grid in the Western Adriatic, and in the Eastern Adriatic it was augmented with chart soundings (**Figure 7**).

In terms of residuals between the check soundings and predicted depth, there is a slight improvement of the RMSE of 5% (RMSE = 13 m).

4.3.2 Comparison with global bathymetric models in the Adriatic Sea

In this section, the GGM+ DBM was compared with data from global bathymetric grids in the Adriatic Sea: DTU10BAT [26], ETOPO 1 [7], GEBCO 2021 [6], SRTM 15+ v2.3 [5], and SS v20.1 [4]. Absolute differences between the GGM+ DBM and global grids at mutual 15 arc seconds resolution are presented in **Figure 8**.

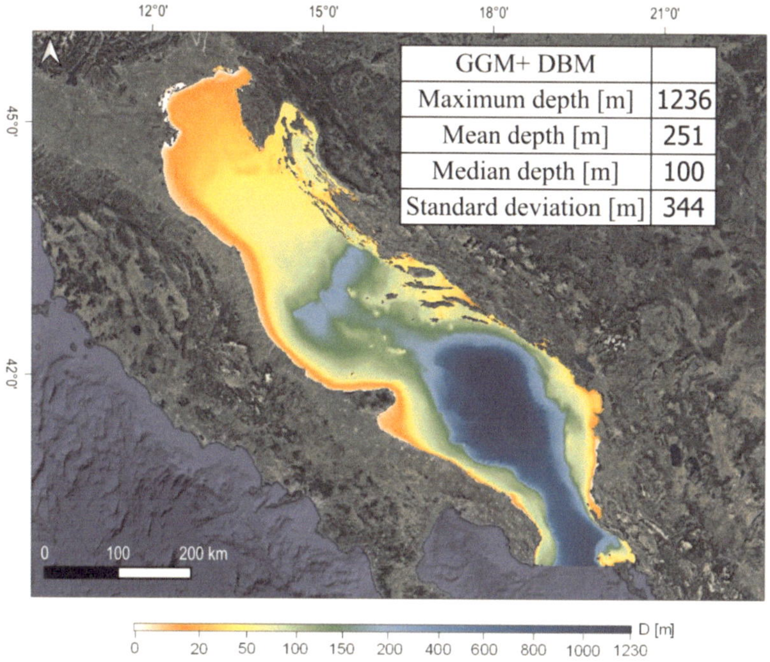

Figure 7.
Bathymetric model of the Adriatic Sea GGM+ DBM.

Statistics of absolute differences between GGM+ DBM and analyzed bathymetry models over study area are given in **Table 4**.

As compared to the GGM+ DBM, the DTU10BAT, and the ETOPO 1, bathymetric models were the models with the greatest discrepancies throughout the study area, especially along the eastern coast. SS v20.1 and SRTM15+ v2.3 showed similar spatial distribution of absolute differences. GEBCO 2021 DBM had the best alignment with the GGM+ DBM over the study area with the median absolute difference of 2 meters. This is less than 1% of the average depth of the Adriatic Sea. Generally, absolute differences along the eastern well-indented coast are larger than along the western coast for all the models. The level of similarity and homogeneity between models is highly influenced by the input data and methodology upon which the grids were constructed, especially the distribution and quality of the input bathymetry.

4.3.3 Quality assessment of bathymetric models in Adriatic Sea

Depth of the analyzed DBMs were compared to the check soundings. The RMSE of the DBMs in the Adriatic Sea is presented in **Table 5**.

More recent digital bathymetric models (GEBCO 2021, GGM+ DBM, SS v20.1, and SRTM 15+ v2.3) showed better accuracy than the older versions (DTU10BAT and ETOPO1). Recent versions were derived from up-to-date altimetry data and/or more dense bathymetry data.

Quality of the DBMs was compared in different depth ranges: 0–20 m, 20–50 m, 50–100 m, 100–200 m, and deeper than 200 m (**Figure 9**). The NRMSE was chosen as a quality measure.

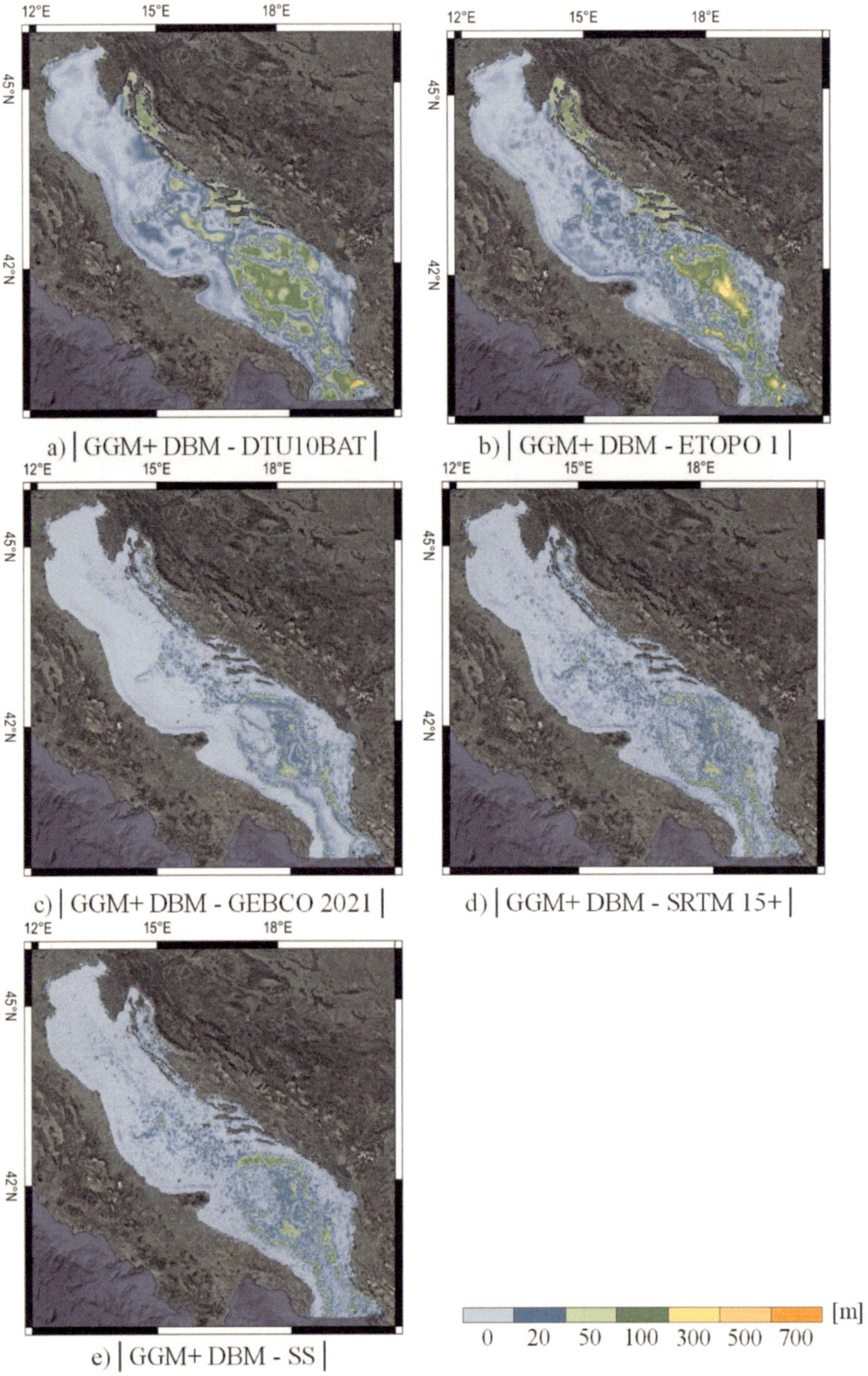

Figure 8.
Absolute difference between GGM+ DBM and (a) DTU10BAT, (b) ETOPO1, (c) GEBCO 2021, (d)SRTM 15+, and (e) SS in the Adriatic.

|ΔD|	MAX [m]	MEAN [m]	σ [m]	MEDIAN [m]
GGM+ DBM - DTU10BAT	686	26	39	8
GGM+ DBM – ETOPO 1	482	26	47	7
GGM+ DBM - GEBCO 2021	214	7	12	2
GGM+ DBM - SRTM15+ v2.3	562	9	18	4
GGM+ DBM - SS v20.1	585	10	19	3

Table 4.
Statistics of absolute differences between GGM+ DBM and global grids in the Adriatic.

DBM	RMSE [m]
DTU10BAT	36
ETOPO1	36
GEBCO 2021	17
SRTM15+ v2.3	11
SS v20.1	16
GGM+ DBM	13

Table 5.
Quality of bathymetric models in the Adriatic Sea.

Generally, the lowest accuracy of the predicted depth was in the shallowest depth range, up to 20 meters deep. The error was larger than 100% of the depth for the SS v20.1 DBM. Lower accuracy is the result of the coarse resolution of the models and the limitation of altimeter technology in coastal areas. As presented in **Figure 9**, bathymetry estimated from altimeter-derived gravity anomalies had better agreement with seafloor topography in deeper seas. In marine areas in the Adriatic Sea that are over 200 meters deep, the accuracy of bathymetric estimation was up to 10% of the depth.

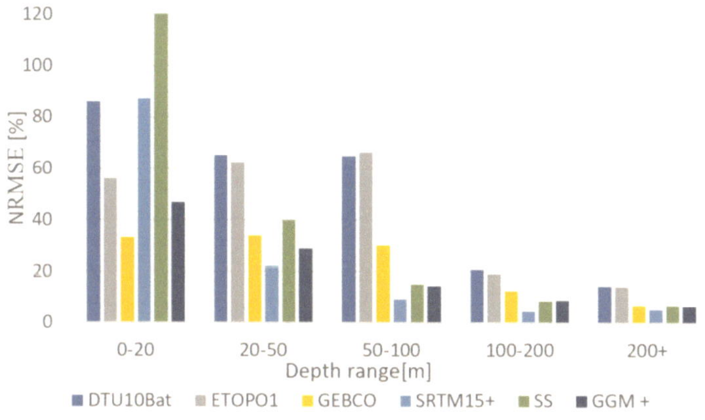

Figure 9.
Quality of bathymetric models in the Adriatic Sea in different depth ranges.

5. Conclusion

Altimeter technology enhanced our knowledge of the seafloor topography and revealed morphological features of the unmapped ocean areas. Widely used global bathymetry models are calculated by the Smith and Sandwell approach (S&S) in the frequency domain. This chapter presented the possibility of regional bathymetry modeling by the gravity-geologic method (GGM) in the space domain with a simpler algorithm, higher resolution, and satisfactory quality as compared to global solutions.

The digital bathymetry model of the Adriatic Sea with 1/16 arc-minute grid spacing (GGM+ DBM) was estimated from the DTU10 model of marine gravity anomalies by the GGM method. Density contrast between seafloor and seawater of 15 000 kgm^{-3}, selected from the trade-off diagram, had minimized the root mean square error (RMSE). The model was augmented by depth soundings from the EMODnet grid in the West Adriatic and nautical charts in the East Adriatic. GGM+ DBM is well adjusted to the topography of the Adriatic Sea, with the RMSE of 13 m.

As compared to modern shipborne bathymetric surveys, bathymetry estimated from altimetry has a coarse spatial resolution and lower accuracy, especially in coastal areas. The greatest discrepancies between the global grids and the GGM+ DBM are along the eastern Adriatic coast due to altimetry limitation and diverse input bathymetry. As compared to chart soundings, all models had the lowest accuracy in the coastal area shallower than 20 m. The quality increased up to 10% of the depth in the deepest parts of the Adriatic. Limitations of the bathymetry estimated from altimetry can be overcome by more available high-quality bathymetry in important coastal areas.

Author details

Ljerka Vrdoljak[1*] and Tomislav Bašić[2]

1 Hydrographic Institute of the Republic of Croatia, Split, Croatia

2 University of Zagreb, Croatia

*Address all correspondence to: ljerka.vrdoljak@hhi.hr

References

[1] The Nippon Foundation-GEBCO. Seabed 2030 Project. Available from: https://seabed2030.org/ [Accessed: June 17, 2022]

[2] EMODnet Bathymetry Consortium: EMODnet Digital Bathymetry (DTM). 2020. DOI: https://doi.org/10.12770/bb6a87dd-e579-4036-abe1-e649cea9881a

[3] International Hydrographic Organization (IHO). Guidance on Crowdsourced Bathymery. Draft Ed 2. 2019; 51 p. Available online at: https://iho.int/iho_pubs/IHO_Download.htm

[4] Smith WHF, Sandwell DT. Global sea floor topography from satellite altimetry and ship depth soundings. Science. 1997; **277**:1956-1962

[5] Tozer B, Sandwell DT, Smith WHF, Olson C, Beale JR, Wessel P. Global bathymetry and topography at 15 arc sec: SRTM15+. Earth and Space Science. 2019;**6**(10):1847-1864

[6] GEBCO Compilation Group: GEBCO 2021 Grid. 2021

[7] Amante C, Eakins BW. ETOPO1 1 Arc-Minute Global Relief Model: Procedures, data sources and analysis. NOAA Technical Memorandum NESDIS NGDC-24. 2009

[8] Dierssen H, Theberge A. Bathymetry: Assessing methods. In: Wang Y, editor. Encyclopedia of Natural Resources II: Water and Air. 1 st ed. Boca Raton: Taylor & Francis Group; 2014. pp. 628-636

[9] Smith WHF, Sandwell DT. Bathymetric prediction from dense satellite altimetry and sparse shipboard bathymetry. Journal of Geophysical Research - Solid Earth. 1994;**99**: 803-821

[10] International Hydrographic Organization (IHO). IHO Data Centre for Digital Bathymetry (DCBC). 2022. Available from: https://www.ngdc.noaa.gov/iho/ [Accessed: June 14, 2022]

[11] Sandwell DT, Smith HF. Bathymetric estimation. In: Fu LL, Cazenave A, editors. Satellite Altimetry and Earth Sciences. San Diego: Academic Press; 2000. pp. 441-457. DOI: 10.1016/S0074-6142(01)80157-1

[12] Calmant S, Baudry N. Modelling bathymetry by inverting satellite altimetry data: A review. Marine Geophysical Researches. 1996;**18**: 123-134. DOI: 10.1007/BF00286073

[13] Parker RL. The rapid calculation of potential anomalies. Geophysical Journal of the Royal Astronomical Society. 1973; **31**:447-455. DOI: 10.1111/j.1365- 246X.1973.tb06513.x

[14] Watts AB. Isostasy and Flexure of the Lithosphere. Cambridge: Cambridge University Press; 2014. p. 458

[15] Ibrahim A, Hinze WJ. Mapping buried bedrock topography with gravity. Ground Water. 1972;**10**(3):18-23. DOI: 10.1111/j.1745-6584.1972.tb02921.x

[16] Kim J, Von Frese R, Lee B, Roman D, Doh SJ. Altimetry-derived gravity predictions of bathymetry by the gravity-geologic method. Pure and Applied Geophysics. 2011;**168**:815-826. DOI: 10.1007/s00024-010-0170-5

[17] Hsiao YS, Cheinway H, Cheng YS, Chen LC, Hsu HJ, Tsai JS, et al. High-resolution depth and coastline over major atolls of South China Sea from satellite altimetry and imagery. Remote Sensing of Environment. 2016;**176**: 69-83. DOI: 10.1016/j.rse.2016.01.016

[18] Xiang X, Wan X, Zhang R, Li Y, Sui X, Wang W. Bathymetry inversion with the gravity-geologic method: A study of long-wavelength gravity modeling based on adaptive mesh. Marine Geodesy. 2017;**40**:329-340. DOI: 10.1080/01490419.2017.1335257

[19] Yeu Y, Yee JJ, Yun HS, Kim BK. Evaluation of the Accuracy of Bathymetry on the Nearshore Coastlines of Western Korea from Satellite Altimetry, Multi-Beam, and Airborne Bathymetric LiDAR. Sensors. 2018;**18**: 2926. DOI: 10.3390/s18092926

[20] Sun Z, Ouyang M, Guan B. Bathymetry predicting using the altimetry gravity anomalies in South China Sea. Geodesy and Geodynamics. 2018;**9**(2):156-161. DOI: 10.1016/j.geog.2017.07.003

[21] DTU Space. DTU1oBAT digital bathymetry model. 2022. Available from: https://www.space.dtu.dk/english/research/scientific_data_and_models [Accessed: June 14, 2022]

[22] NOAA. ETOPO 1 digital bathymetry model. 2022. Available from: https://www.ngdc.noaa.gov/mgg/global/ [Accessed: June 14, 2022]

[23] GEBCO. GEBCO 2021 digital bathymetry model. 2022. Available from: https://download.gebco.net/ [Accessed: June 14, 2022]

[24] Scripps Institute of Oceanography. SRTM15+ v.2.3 digital bathymetry model. 2022. Available from: https://topex.ucsd.edu/pub/archive/srtm15/ [Accessed: June 14, 2022]

[25] Scripps Institute of Oceanography. SS v.23.1 digital bathymetry model. 2022. Available from: https://topex.ucsd.edu/pub/global_topo_1min/ [Accessed: June 14, 2022]

[26] Andersen OB. The DTU10 Gravity field and Mean Sea surface. In: Presentation on Second International Symposium of the Gravity Field of the Earth (IGFS2). Alaska; 2010

[27] Marks KM, Smith WHF. An evaluation of publicly available bathymetry grid. Marine Geophysical Researches. 2006;**27**:19-34. DOI: 10.1007/s11001-005-2095-4

[28] Hao R, Wan X, Wang Y, Annan RF. Evaluation of four global bathymetry models by shipborne depths data. Journal of Surveying Engineering. 2021; **148**:2. DOI: 10.1061/(ASCE)SU.1943-5428.0000392

[29] Abramova AS. Comparison of publicly available global bathymetry grids. In: Presentation on Gebco Fifth Science Day. Callao, Peru; 2010

[30] Lj V. Comparison and analysis of publicly available bathymetry models in the East Adriatic Sea. Naše more. 2021;**2**: 110-119. DOI: 10.17818/nm/2021/2.7

[31] Sandwell DT, Muller RD, Smith WH, Garcia E, Francis R. New global marine gravity model from CryoSat-2 and Jason-1 reveals buried tectonic structure. Science. 2014;**346**(65):65-67. DOI: 10.1126/science.1258213

[32] Roman DR. An Integrated Geophysical Investigation of Greenland's Tectonic History. Columbus: Ohio State University; 1999

[33] Wei Z, Jinyun G, Zhu C, Yuan J, Chang X, Ji B. Evaluating accuracy of HY-2A/GM-derived gravity data with the gravity-geologic method to predict bathymetry. Frontiers in Earth Science. 2021;**9**:636246

[34] International Hydrographic Organization (IHO). Limits of Oceans and Seas. 3rd ed. Monaco; 1953. p. 38

[35] Flanders Marine Institute. IHO Sea Areas, version 3. 2018. Available from: https://www.marineregions.org/

[36] Vrdoljak L, Režić M, Petričević I. Bathymetric and Geological Properties of the Adriatic Sea. Rudarsko-geološko-naftni zbornik. 2021;**36**(2):93-107. DOI: 10.17794/rgn.2021.2.9

[37] Intergovernmental Oceanographic Commission IOC, International Hydrographic Organization IHO, British Oceanographic Data Centre BODC. Centenary Edition of the GEBCO Digital Atlas. Liverpool: Published on CD-ROM; 2003

Chapter 5

Terrain Corrections in Gravity and Gradiometry

Sajjad Sajjadi and Zdenek Martinec

Abstract

Before the computation of short-wavelength and long-wavelength components of the geoid undulations from terrestrial data and the two latest satellite missions, i.e. gravity (GRACE mission) and gradiometry (GOCE mission) measurements, the terrain corrections must be determined. Since the corrections enter the first of the three steps of the Remove-Compute-Restore (RCR) procedure for applying Stokes's integral, this study focuses on determining these corrections. Formulation of the effects introduced and the effects are computed over high elevated topography in Ireland using Helmert's second condensation method. Finally, the effects of topography on geoid height determinations are presented.

Keywords: gravity, gradiometry, terrain correction, Remove-Compute-Restore procedure (RCR), topographical effects

1. Introduction

The geoid is defined as an equipotential surface along which the Earth's gravity potential (W) is constant and equal to a reference value W_0. This datum is chosen such that the geoid coincides with a mean level of the oceans and can be mathematically extended over the continents. As a result of the unequal distribution of masses in the Earth's interior, the geoid is irregularly shaped. It describes the figure of the Earth by a physical quantity, the gravity potential, in contrast to the idealized geometrical figure of a reference *ellipsoid*. The separations between the two surfaces are called the geoid *undulation* N, or geoidal heights.

The local gravity potential, W_{local}, value is derived from gravimetric (N_{grav}), and geometric (N_{geo}) geoid undulations. The geometric geoid undulations are obtained from GNSS and leveling data. In contrast, the gravimetric geoid undulations are computed based upon their long-wavelength components from a Global Geopotential Model (GGM), and their short-wavelength components using the terrestrial data through the so-called Remove-Compute-Restore (RCR) approach from gravity measurements.[1]

[1] **Gravimetry** is the method of measuring gravity and the instrument used is called a *gravimeter*. In the past, gravity data were exclusively provided by terrestrial surveys. Later, transportable relative gravimeters were designed for the use on ship and airborne, however, the data accuracy was very variable and geographically unevenly distributed. Recently, satellites gravimetry has emerged providing the global coverage of repeated measurements. The unit of gravity is the *Galileo*, 1 mGal $= 10^{-5}$ m.s^{-2}.

1.1 The Remove-Compute-Restore procedure (RCR)

The RCR procedure is a method that fulfils Stokes' requirements[2] for computing the geoid from:

- Terrestrial gravity measurements or,
- Satellites gradiometry[3] measurements (e.g. GRACE, GOCE).

A gravimeter measures magnitude of gravity,

$$g = |\text{grad } V|, \tag{1}$$

whereas the gradiometer measures the components of the gradient of gravity. We will focus on the *rr* component of the gradiometric tensor,

$$V_{rr} = (\text{grad grad } V)_{rr}. \tag{2}$$

1.1.1 *Removing the gravitational effect of the residual topographical masses*

Subtracting the gravitational effect of the residual topographical masses, δV, from the actual anomalous gravitational potential T creates potential T^h that is harmonic outside the geoid,

$$T^h = T - \delta V. \tag{3}$$

The gravity attraction of the residual topographical masses is then

$$\delta A := \frac{\partial \delta V}{\partial r} \tag{4}$$

at the point of the gravity measurements, and

$$\delta E := \frac{\partial^2 \delta V}{\partial r^2} \tag{5}$$

at point of gradiometric measurements.

To make a potential harmonic in a space above the geoid, these effects have to be calculated and removed from the observations:

$$\Delta g^h = \Delta g^{obs} - \delta A \tag{6}$$

$$\Delta V^h_{rr} = \Delta V^{obs}_{rr} - \delta E. \tag{7}$$

1.1.2 *Computation of the residual geoid (co-geoid)*

The first step of the computation is continuing Δg^h and ΔV^h_{rr} from the surface or satellite elevation to the geoid. This is usually performed by a harmonic downward continuation (DWC) method [1].

---

[2] No masses outside the geoid, and the measurements are referred to the geoid.

[3] Gravity **gradiometry** is the study of variations in the acceleration due to gravity. It is the measurement of the rate of change of gravitational acceleration called gravity *gradient* is the spatial. The unit of gradient is the *Eötvös*, 1 E= 10^{-9}m/s^2/m.

Assuming that the data have been continued down to the geoid, the two basic requirements are met and the computation of the geoidal heights can now be carried out. However, the computation gives the height N^h of the equipotential surface of T^h that is called *co-geoid*.

$$N^h = \frac{R}{4\pi\gamma_Q} \int_{\Omega_0} \Delta g^h S(\psi) d\Omega, \tag{8}$$

where ψ is the spherical solid angle between the computation point and an integration point and $S(\psi)$ is the Stokes's function ([2], p. 104) integrated over the full solid angle Ω_0

$$S(\psi) = \frac{1}{\sin\frac{\psi}{2}} - 6\sin\frac{\psi}{2} + 1 - 5\cos\psi - 3\cos\psi \ln\left(\sin\frac{\psi}{2} + \sin^2\frac{\psi}{2}\right). \tag{9}$$

Likewise, the heights of the co-geoid can be determined from gradiometric data as

$$N^h = \frac{R^2}{4\pi\gamma_Q} \int_{\Omega_0} \Delta V^h_{rr} K_{rr}(\psi) d\Omega, \tag{10}$$

where the kernel $K_{rr}(\psi)$ is the Green's function as given by Martinec [3]

$$K_{rr}(\psi) = -3 + 6\sin\frac{\psi}{2} + (1 - 3\cos\psi)\ln\left(\frac{\sin\frac{\psi}{2}}{1 + \sin\frac{\psi}{2}}\right). \tag{11}$$

1.1.3 Adding the contribution of the topography to the solution

Finally, the heights of the geoid will be obtained by adding the difference δN shown in **Figure 1** which is called the *primary indirect effect on geoid*.

To find the expression for δN, we start with geoidal height N is derived from Bruns' formula

$$N = \frac{T}{\gamma_Q}, \tag{12}$$

where γ_Q is the normal gravity on a reference ellipsoid. Substituting for T from Eq. (60) we obtain Eq. (3)

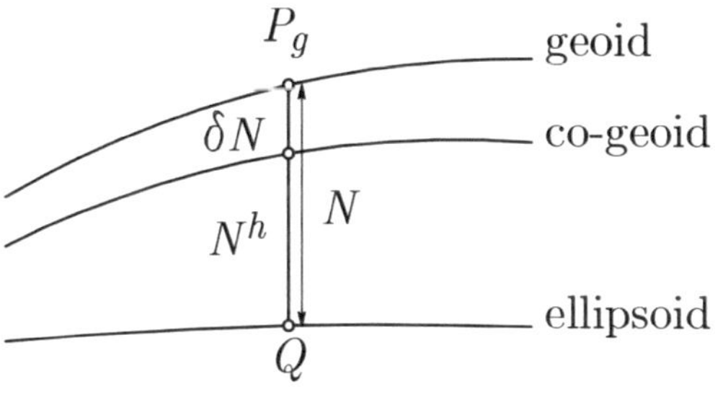

Figure 1.
co-geoid and geoidal heights.

$$N = \frac{T^h + \delta V}{\gamma_Q}. \tag{13}$$

Applying Bruns' formula to the undulation of the co-geoid, $N^h = T^h/\gamma_Q$, the geoidal height N is

$$N = N^h + \delta N, \tag{14}$$

where

$$\delta N = \frac{\delta V}{\gamma_Q}, \tag{15}$$

and δV must be taken at a point on the geoid P_g.

GRACE and GOCE are the two latest satellite missions for precise, long-wavelength geoid determination from gravity and gradiometry measurements, respectively. Prior processing GRACE and GOCE data, the terrain corrections have to be determined.

The RCR procedure for gravity and gravimetry is summarized in **Figure 2**.

It shows the motivation for correcting the data by the effects of the topographical masses. These are the direct topographical effects δA computed at surface or satellite

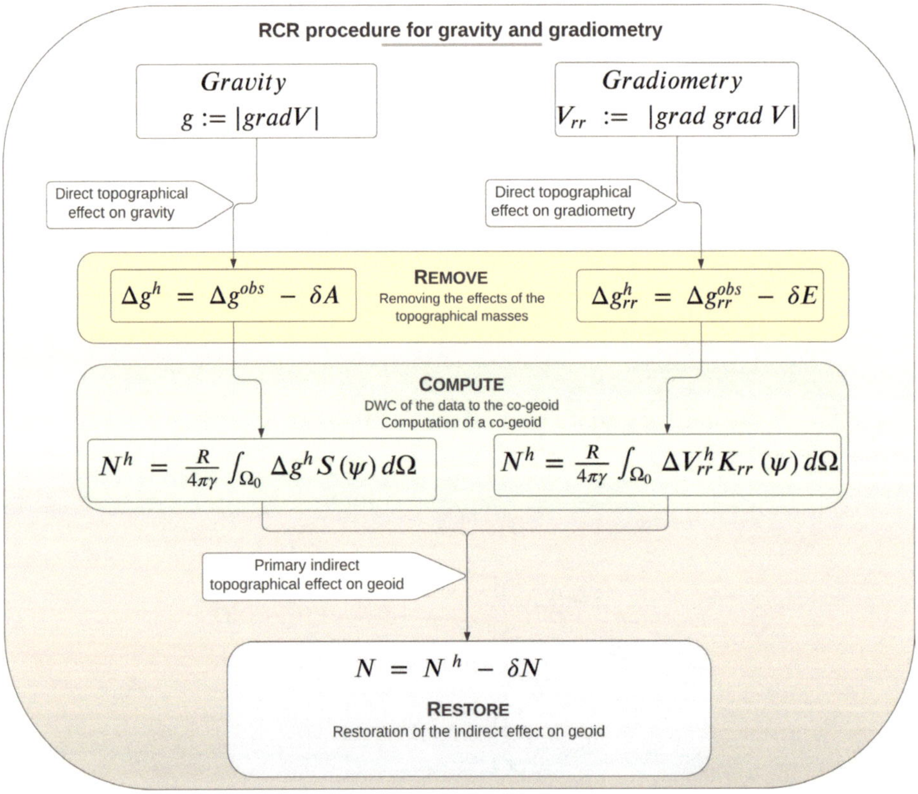

Figure 2.
Summary of the RCR procedure for gravity and gradiometry.

altitude, the direct topographical effect on gradiometry δE computed at satellite altitude and the indirect topographical effect on geoid δN computed on the geoid.

In the intervening period of the theory of physical geodesy, a wide range of methods have been developed, e.g. Airy-isostatic reduction method [4], Residual Terrain Model (RTM) scheme [5], Helmert's first and second method of condensation [4, 6–8]. The choice of each method is area-dependent, and the correlation between the spatial resolution of DEMs, and the elevation of computation points or data surrounding the computation points show the most suitable spatial resolution of DEMs that provide intended geodetic accuracies. This is investigated numerically over high elevated topography in Ireland using Helmert's second condensation method.

The next section is devoted to express them mathematically and prepare them for numerical realization.

2. Topographical effects

The term **terrain effect** is used to express the gravitational effects of topographical masses on gravity anomalies, deflection of the vertical and other observed quantity. It can be classified according to a location of anomalous masses. *Topographical effects* are the direct influence of the visible topography in mountainous areas; *isostatical effects* account for a hypothesized isostatic compensation, whilst the *residual terrain model* (RTM) effects account for short-wavelength topographic irregularities referring topographic elevations to a smooth mean elevation surface, which may be defined, for instance, by spherical harmonic expansion of topographic heights.

2.1 Topographical masses and the Bouguer plate

2.1.1 Topographical masses

The topographical masses are the masses outside the geoid and below the topographical surface. The gravitational potential V^t generated by the topographical masses is

$$V^t(r,\Omega) = G\int_{\Omega_0}\int_{r'=r_g(\Omega')}^{r_t(\Omega')} \frac{\varrho(r',\Omega')}{L(r,\psi,r')} r'^2 dr' d\Omega'. \tag{16}$$

where G is the Newton's gravitational constant, $G = 6.67 \cdot 10^{-11}\text{m}^3.\text{kg}^{-1}.\text{s}^{-2}$, $\varrho(r',\Omega')$ is the mass density inside the Earth's interior located at $P'(r',\Omega'), L(r,\psi,r')$ is the distance between P and P' and ψ is the angular distance between the geocentric directions $\Omega = (\vartheta,\varphi)$ and $\Omega' - (\vartheta',\varphi')$, see **Figure 3**, i.e. $\psi(\Omega,\Omega')$ is the spherical solid angle between P and P',

$$\cos\psi = \cos\vartheta\cos\vartheta' + \sin\vartheta\sin\vartheta'\cos(\varphi-\varphi'). \tag{17}$$

The argument notation in $L(r,\psi,r')$ is used to emphasize the fact that L depends on radial distances r and r', and the angular distance ψ.

$$L(r,\psi,r') := \sqrt{r'^2 - 2rr'\cos\psi + r^2}. \tag{18}$$

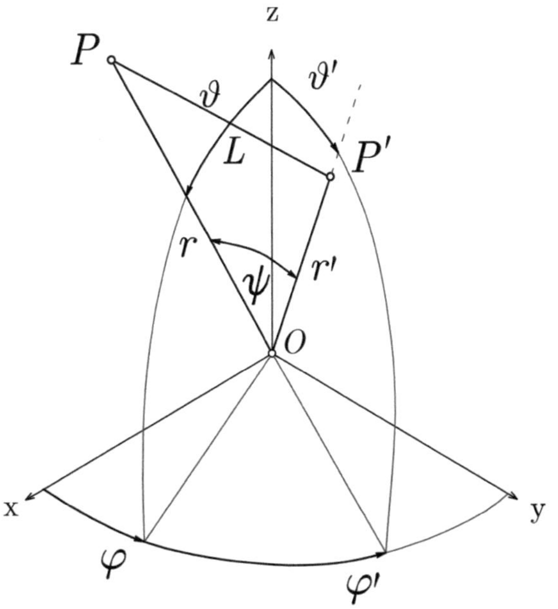

Figure 3.
Spherical coordinates of the computation point $P(r,\Omega)$, an integration point $P'(r',\Omega')$, the distance L and solid angle ψ between P and P'.

To abbreviate notations, we introduced the symbol $\tilde{L}^{-1}(r,\psi,r')$ for an indefinite radial integral of the Newton kernel,

$$\tilde{L}^{-1}(r,\psi,r') := \int_{r'} \frac{r'^2}{L(r,\psi,r')} dr'. \tag{19}$$

Assuming that the density of the topographical masses does not vary in radial direction, that is $\varrho(r',\Omega') = \varrho(\Omega')$, and substituting Eq. (19) in Eq. (16), the Newton's volume integral for the gravitational potential V^t becomes

$$V^t(r,\Omega) = G\int_{\Omega_0} \varrho(\Omega')\tilde{L}^{-1}(r,\psi,r')\Big|_{r'=r_g(\Omega')}^{r_t(\Omega')} d\Omega'. \tag{20}$$

2.1.2 Bouguer plate

The Bouguer plate, used as an approximate model in gravity and gravity anomaly computations accounts for the bulk of topographical effects.

In Cartesian geometry **Figure 4a**, the topography around the gravity station P is approximated by an infinite plate of thickness H_P and the masses between the geoid and the Earth's surface have a constant density ϱ equal to mean topographical density $\varrho_0 = 2670\,\text{kg.m}^{-3}$.

In spherical geometry **Figure 4b**, the Bouguer plate is regarded as a spherical layer of thickness H_P and density ϱ_0. The gravitational potential of the spherical Bouguer layer is

$$V^B(r,\Omega) = G\varrho_0\int_{\Omega_0} \tilde{L}^{-1}(r,\psi,r')\Big|_{r'=r_g(\Omega)}^{r_t(\Omega)} d\Omega'. \tag{21}$$

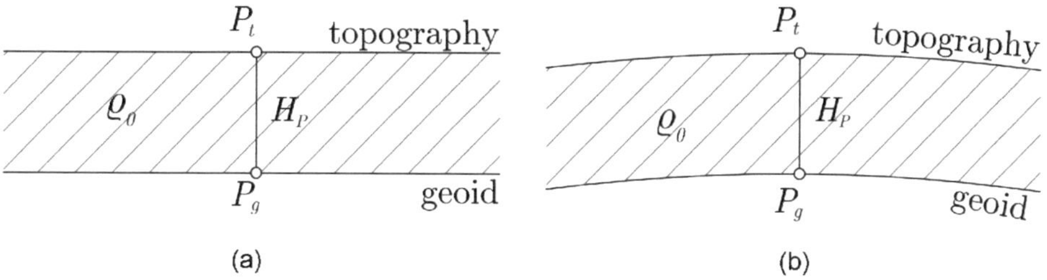

Figure 4.
The Bouguer plate in Cartesian and spherical geometry. (a) Infinite Bouguer plate. (b) Spherical Bouguer layer.

For evaluating this integral, the geoid is approximated by a sphere of radius R

$$r_g(\Omega) = R \quad \text{so} \quad r_t(\Omega) = R + H(\Omega). \tag{22}$$

For R we consider the mean radius of the Earth, R = 6371 km. The integral Eq. (21) can be evaluated analytically, e.g. Wichiencharoen [9],

$$V^B(r,\Omega) = \begin{cases} 4\pi G\varrho_0 \frac{1}{r}\left[R^2 H(\Omega) + RH^2(\Omega) + \frac{1}{3}H^3(\Omega)\right] & , r \geq r_t(\Omega), \\ 2\pi G\varrho_0 \left[(R+H(\Omega))^2 - \frac{2}{3}\frac{R^3}{r} - \frac{1}{3}r^2\right] & , R \leq r \leq r_t(\Omega), \\ 4\pi G\varrho_0 \left[RH(\Omega) + \frac{1}{2}H^2(\Omega)\right] & , r \leq R. \end{cases} \tag{23}$$

2.1.3 Terrain roughness

Since the actual Earth's surface deviates from the Bouguer sphere, there are deficiencies and abundances of topographical masses with respect to the mass of the Bouguer plate **Figure 5**. These contribute to the topographical potential V^t through the term V^R as

$$V^t(r,\Omega) = V^B(r,\Omega) + V^R(r,\Omega). \tag{24}$$

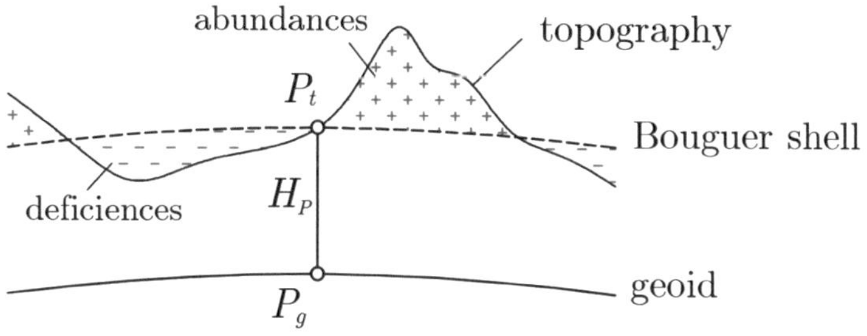

Figure 5.
Roughness of the terrain.

The *terrain roughness term* V^R is expressed by the Newton integral

$$V^R(r,\Omega) = G\varrho_0 \int_{\Omega_0} \left[\tilde{L}^{-1}(r,\psi,r')\Big|_{r'=R}^{r_t(\Omega')} - \tilde{L}^{-1}(r,\psi,r')\Big|_{r'=R}^{r_t(\Omega)} \right] d\Omega'. \tag{25}$$

2.2 Compensated masses and the Helmert condensation layer

2.2.1 Compensation of the gravitational effects of topographical masses

The equipotential surfaces of V^t undulate by several hundreds of meters with respect to a level ellipsoid. The fact that the known undulations of the geoid are much smaller than those induced by potential V^t indicates that there must exist a compensation mechanism which reduces the gravitational effect of topographical masses [10].

As the masses are compensated in some way [2], we can introduce the *gravitational potential of compensated masses* V^c as an approximation of the topographical potential V^t. The difference between V^t and V^c is called *the residual topographical potential* δV:

$$\delta V := V^t - V^c. \tag{26}$$

Two extremely idealized *isostatic compensation models* (see **Figure 6**) were proposed to approximate the effect of topographical abundances from surface gravity observations.

The *Pratt-Hayford model* was outlined by J.H Pratt in 1854 and put into a mathematical form by J.F Hayford. According to Pratt, the mountains have risen from the underground somewhat like a fermenting dough [2, sect.3]. The topographical masses are compensated by varying density distribution within the layer of a constant thickness D = 100 km under the geoid and a density $\varrho_c(\Omega)$.

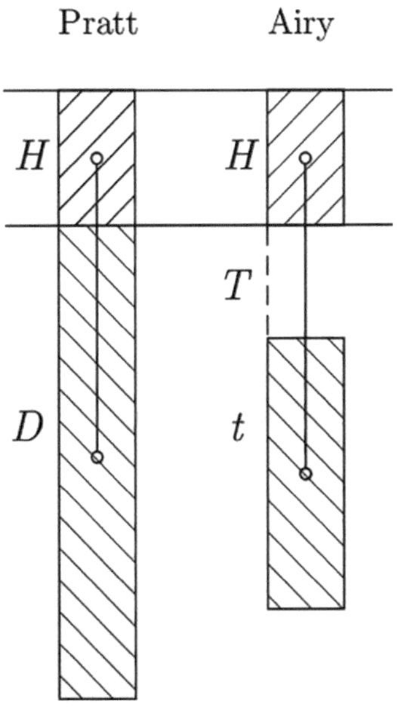

Figure 6.
topography and compensation layers of Pratt-Hayford and Airy-Heiskanen models ([2], Fig 3.16).

The *Airy-heiskanen model*, proposed by G.B Airy in 1855 and formulated for geodetic purposes by W.A Heiskanen, assumes that the mountains are floating on a fluid lava of higher known density ϱ_1 (somewhat like an iceberg floating on water), so that the higher the mountain, the deeper it sinks [2, sect.3]. The topographical masses are compensated by varying thickness $t(\Omega)$ of a compensation layer which surface is situated at $T = 30$ km deep. The density of the compensation layer is considered constant and equal to the density difference $\varrho_c = \varrho_1 - \varrho_0(>0)$.

2.2.2 Helmert condensation layer

In the limiting case, the topographical masses may be compensated by a thin mass layer located on the geoid (somewhat like a glass sphere made over very thin but very robust glass [2]). As shown **Figure 7**, the topographical masses are condensed as a surface mass layer on the geoid. This kind of compensation is called *Helmert 2nd condensation* [11] approximating the actual potential of the topographical masses V^t by the potential of a single layer V^c described by Newton's surface integral as

$$V^c(r,\Omega) = GR^2\int_{\Omega_0} \sigma(\Omega')L^{-1}(r,\psi,R)d\Omega', \tag{27}$$

where $\sigma(\Omega)$ is a surface density of Helmert's condensation layer, and L^{-1} is the reciprocal distance $1/L$. Analogously to Eq. (24), we can rewrite Eq. (27) as

$$V^c(r,\Omega) = V^{\sigma,B}(r,\Omega) + V^{\sigma,R}(r,\Omega), \tag{28}$$

where

$$V^{\sigma,R}(r,\Omega) = GR^2\int_{\Omega_0} [\sigma(\Omega') - \sigma(\Omega)]L^{-1}(r,\psi,R)d\Omega'. \tag{29}$$

The symbol $V^{\sigma,B}(r,\Omega)$ denotes the gravitational potential of a spherical layer with density $\sigma(\Omega)$ and radius R,

$$V^{\sigma,B}(r,\Omega) := GR^2\sigma(\Omega)\int_{\Omega_0} L^{-1}(r,\psi,R)d\Omega', \tag{30}$$

that may be evaluated analytically by

$$V^{\sigma,B}(r,\Omega) = \begin{cases} 4\pi G\sigma(\Omega)\dfrac{R^2}{r} & ,r>R, \\ 4\pi G\sigma(\Omega)R & ,r\le R. \end{cases} \tag{31}$$

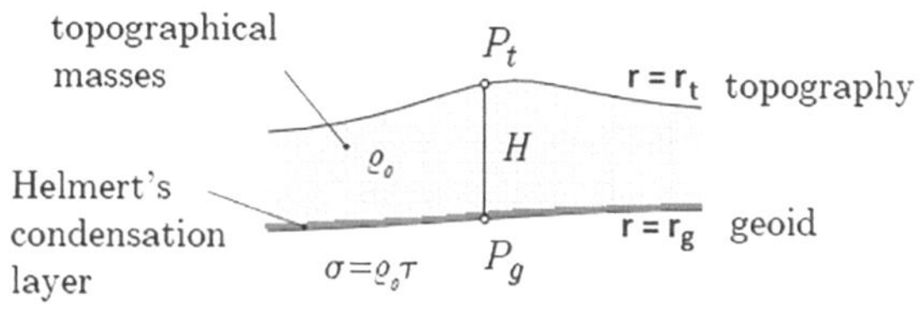

Figure 7.
Helmert condensation layer of density σ.

The condensation density $\sigma(\Omega)$ can be chosen in a variety of ways. For this study, it will be chosen as in Martinec and Vaníček and Martinec [10] such that the principle of conservation of topographical masses [9] is respected. Assuming that

$$V^B(r,\Omega) = V^{\sigma,B}(r,\Omega) \qquad \text{for } r = r_t(\Omega), \tag{32}$$

and substituting Eqs.(23) and (31) into (32), we find

$$\sigma(\Omega) = \varrho_0 \tau(\Omega), \tag{33}$$

with

$$\tau(\Omega) = H(\Omega)\left(1 + \frac{H(\Omega)}{R} + \frac{H^2(\Omega)}{3R^2}\right). \tag{34}$$

Thus, Eq. (29) will be written

$$V^{\sigma,R}(r,\Omega) = G\varrho_0 R^2 \int_{\Omega_0} [\tau(\Omega') - \tau(\Omega)] L^{-1}(r,\psi,R)\, d\Omega'. \tag{35}$$

2.3 Indirect topographical effect

2.3.1 Indirect topographical effect on potential

The primary indirect topographical effect is the residual potential $\delta V = V^t - V^c$ evaluated at a point $P_g(R,\Omega)$ on the geoid. Considering Eq. (24) for the topographical potential V^t, Eq. (28) for the condensation potential V^c and replacing r by R, we may split δV into two terms as

$$\delta V(R,\Omega) = \delta V^B(R,\Omega) + \delta V^R(R,\Omega), \tag{36}$$

where the Bouguer term and the terrain roughness term are respectively given by

$$\delta V^B(R,\Omega) = V^B(R,\Omega) - V^{\sigma,B}(R,\Omega), \tag{37}$$

$$\delta V^R(R,\Omega) = V^R(R,\Omega) - V^{\sigma,R}(R,\Omega). \tag{38}$$

The subtraction of Eqs. (23) and (31) at $r \leq R$ leads to

$$\delta V^B(R,\Omega) = -2\pi G\varrho_0 H(\Omega)\left(1 + \frac{2}{3}\frac{H(\Omega)}{R}\right), \tag{39}$$

and the subtraction of Eqs. (25) and (35) gives

$$\delta V^R(R,\Omega) = G\varrho_0 \int_{\Omega_0} \Big[\tilde{L}^{-1}(R,\psi,r')\Big|_{r'=R}^{r_t(\Omega')} - \tilde{L}^{-1}(R,\psi,r')\Big|_{r'=R}^{r_t(\Omega)} \\ - R^2[\tau(\Omega') - \tau(\Omega)] L^{-1}(R,\psi,R)\Big] d\Omega'. \tag{40}$$

Therefore, substituting Eqs. (39) and (40) in Eq. (36), we obtain the expression of the primary topographical indirect effect on potential:

$$\delta V(R,\Omega) = -2\pi G\varrho_0 H^2(\Omega)\left(1+\frac{2}{3}\frac{H(\Omega)}{R}\right) + G\varrho_0\int_{\Omega_0}\left[\tilde{L}^{-1}(R,\psi,r')\Big|_{r'=R}^{r_t(\Omega')} - \tilde{L}^{-1}(R,\psi,r')\Big|_{r'=R}^{r_t(\Omega)} - R^2[\tau(\Omega')-\tau(\Omega)]L^{-1}(R,\psi,R)\right]d\Omega'. \tag{41}$$

The unit of this effect is $m^2.s^{-2}$.

2.3.2 Indirect topographical effect on the geoid

To correct the geoidal heights N by this effect, δV is divided by the normal gravity γ_Q (see Eq. (15)),

$$\delta N(R,\Omega) = \frac{\delta V(R,\Omega)}{\gamma_Q}, \tag{42}$$

which is the primary indirect effect on the geoid. Where the normal gravity can be taken as the mean value of the gravity of the Earth $\gamma_Q = 9.81\ m.s^{-2}$. The unit of δN is meters.

2.3.3 The secondary indirect topographical effect (SITE) on gravity

This effect is expressed by means of the Primary Indirect Topographical Effect (PITE) on gravity, δV_{P_g}, at a point on the geoid multiplied by $2/R$,

$$\delta S(\Omega) = \frac{2}{R}\delta V_{P_g}(\Omega). \tag{43}$$

The unit of this effect is mGal.

Notice that the radial derivative of the Newton surface and volume integrals, which is required for computing $\delta A(\Omega)$, is

$$\frac{\partial \tilde{L}^{-1}(r,\psi,r')}{\partial r} = \int_{r'} r'^2 \frac{\partial L^{-1}(r,\psi,r')}{\partial r}dr', \tag{44}$$

and using the derivation of $(u^\alpha)' = \alpha u' u^{\alpha-1}$ in Eq. (18),

$$\frac{\partial L^{-1}(r,\psi,r')}{\partial r} = -\frac{r - r'\cos\psi}{\left(r'^2 - 2rr'\cos\psi + r^2\right)^{3/2}}, \tag{45}$$

where substituting the expression (45) in (44), the analytical expression is given by,

$$\frac{\partial \tilde{L}^{-1}(r,\psi,r')}{\partial r} = \left(3r^2\cos\psi + r\left(r' - 6\cos\psi^2 r'\right) + \cos\psi r'^2\right)L^{-1}(r,\psi,r') + r\left(3\cos\psi^2 - 1\right)\ln|L(r,\psi,r') - r\cos\psi + r'| + C. \tag{46}$$

2.4 Direct topographical effect (DTE)

2.4.1 Direct topographical effect on surface gravity

Differentiating the residual topographical potential δV with respect to r and evaluating the result at the point on the topography $P_t(r_t(\Omega), \Omega)$, we obtain the gravitational attraction caused by the direct topographical effect on surface gravity:

$$\delta A(r, \Omega) = \left. \frac{\partial \delta V(r, \Omega)}{\partial r} \right|_{r=r_t(\Omega)}. \tag{47}$$

Substituting for δV from Eq. (26), we can write

$$\delta A(r, \Omega) = A^t(r, \Omega) - A^c(r, \Omega), \tag{48}$$

where

$$A^t(r, \Omega) = \left. \frac{\partial V^t(r, \Omega)}{\partial r} \right|_{r=r_t(\Omega)} \quad \text{and} \quad A^c(r, \Omega) = \left. \frac{\partial V^c(r, \Omega)}{\partial r} \right|_{r=r_t(\Omega)}, \tag{49}$$

are the radial components of the gravitational attraction induced by the topographical and compensated masses a the point on the Earth's surface, respectively. Considering Eq. (24) for A^t and Eq. (28) for A^c, the attraction change δA may be split into two terms

$$\delta A(r, \Omega) = \delta A^B(r, \Omega) + \delta A^R(r, \Omega), \tag{50}$$

where δA^B represents the *Bouguer term* and δA^R the *terrain roughness term*. Let us start with the determination of the Bouguer term

$$\delta A^B(r, \Omega) = A^B(r, \Omega) - A^{\sigma,B}(r, \Omega), \tag{51}$$

where

$$A^B(r, \Omega) = \left. \frac{\partial V^B(r, \Omega)}{\partial r} \right|_{r=r_t(\Omega)} \quad \text{and} \quad A^{\sigma,B}(r, \Omega) = \left. \frac{\partial V^{\sigma,B}(r, \Omega)}{\partial r} \right|_{r=r_t(\Omega)}. \tag{52}$$

Subsituting the radial derivative of Eq. (23) at $r \geq r_t(\Omega)$ for A^B

$$\frac{\partial V^B(r, \Omega)}{\partial r} = -4\pi G \varrho_0 \frac{1}{r^2} \left[R^2 H(\Omega) + R H^2(\Omega) + \frac{1}{3} H^3(\Omega) \right], \tag{53}$$

similarly, the radial derivative of Eq. (31) at $r > R$ for $A^{\sigma,B}$

$$\frac{\partial V^{\sigma,B}(r, \Omega)}{\partial r} = -4\pi G \sigma(\Omega) \frac{R^2}{r^2}. \tag{54}$$

Using Eqs. (33)–(34) for $\sigma(\Omega)$, we obtain $\delta A^B(r, \Omega) = 0$. Consequently, the direct topographical effect δA only consists of the terrain roughness contribution δA^R that is

$$\delta A^R(r,\Omega) = A^R(r,\Omega) - A^{\sigma,R}(r,\Omega), \tag{55}$$

where

$$A^R(r,\Omega) = \left.\frac{\partial V^R(r,\Omega)}{\partial r}\right|_{r=r_t(\Omega)} \quad \text{and} \quad A^{\sigma,R}(r,\Omega) = \left.\frac{\partial V^{\sigma,R}(r,\Omega)}{\partial r}\right|_{r=r_t(\Omega)}. \tag{56}$$

Substituting the radial derivations of Eq. (25) for A^R and Eq. (35) for $A^{\sigma,R}$ into Eq. (55), we obtain the expression of the direct topographical effect on surface gravity as

$$\delta A(r,\Omega) = G\varrho_0 \int_{\Omega_0} \left[-\left.\frac{\partial \tilde{L}^{-1}(r,\psi,r')}{\partial r}\right|_{r'=R}^{r_t(\Omega)} - R^2[\tau(\Omega') - \tau(\Omega)] \frac{\partial L^{-1}(r,\psi,R)}{\partial r} \right]_{r=r_t(\Omega)} d\Omega'. \tag{57}$$

2.4.2 Direct topographical effect on satellite gravity

Differentiating δV with respect to r and evaluating the result at the point of measurement on satellite $(r = r_{sat}(\Omega), \Omega)$, we obtain the change of the gravitational attraction caused by the direct topographical effect on satellite gravity

$$\delta A(r,\Omega) = \left.\frac{\partial \delta V(r,\Omega)}{\partial r}\right|_{r=r_{sat}(\Omega)}. \tag{58}$$

Analogically to the direct topographical effect on surface gravity Eq. (57), where the radius of the computation point is $r = r_t(\Omega)$, the direct topographical effect on satellite gravity will be:

$$\delta A(r,\Omega) = G\varrho_0 \int_{\Omega_0} \left[\left.\frac{\partial \tilde{L}^{-1}(r,\psi,r')}{\partial r}\right|_{r'=R}^{r_t(\Omega')} - \left.\frac{\partial \tilde{L}^{-1}(r,\psi,r')}{\partial r}\right|_{r'=R}^{r_t(\Omega)} - R^2[\tau(\Omega') - \tau(\Omega)] \frac{\partial L^{-1}(r,\psi,R)}{\partial r} \right]_{r=r_{sat}(\Omega)} d\Omega', \tag{59}$$

where $r_{sat}(\Omega)$, is:

$$r_{sat}(\Omega) = R + H_{sat}, \tag{60}$$

and H_{sat} is the flying altitude of the satellite[4] which performs the gravity measurements. For instance, the value of $H_{sat}(\Omega)$ for the satellite GRACE is $H_{sat}(\Omega) =$ 400 km.

[4] Note that the same formula applies to air-borne gravity measurements, replacing the radius of computation by $r = R + H_{plane}$ where H_{plane} is the flying altitude of the plane performing gravity measurement.

2.4.3 Direct topographical effect on gradiometry

Differentiating δV twice with respect to r and evaluating the result at the point of measurement on satellite $(r = r_{sat}(\Omega), \Omega)$, we obtain the rr component of the gradient of gravity caused by the direct topographical effect on gradiometry and Eq. (5) becomes

$$\delta E(r, \Omega) = \left. \frac{\partial^2 \delta V(r, \Omega)}{\partial r^2} \right|_{r=r_{sat}(\Omega)}. \tag{61}$$

Substituting for the residual topographical potential δV from Eq. (26), we can write

$$\delta E(r, \Omega) = V^t_{rr}(r, \Omega) - V^c_{rr}(r, \Omega), \tag{62}$$

where

$$V^t_{rr}(r, \Omega) = \left. \frac{\partial^2 V^t(r, \Omega)}{\partial r^2} \right|_{r=r_{sat}(\Omega)} \quad \text{and} \quad V^c_{rr}(r, \Omega) = \left. \frac{\partial^2 V^c(r, \Omega)}{\partial r^2} \right|_{r=r_{sat}(\Omega)}, \tag{63}$$

are the rr components of the gradiometric tensor induced by the topographical and compensated masses a the point at satellite altitude, respectively. Considering Eq. (24) for V^t_{rr} and Eq. (28) for V^c_{rr}, δE may be split into two terms

$$\delta E(r, \Omega) = \delta E^B(r, \Omega) + \delta E^R(r, \Omega), \tag{64}$$

where δE^B represents the *Bouguer term* and δE^R the *terrain roughness term*.
Let us start with the determination of the Bouguer term

$$\delta E^B(r, \Omega) = V^B_{rr}(r, \Omega) - V^{\sigma,B}_{rr}(r, \Omega), \tag{65}$$

where

$$V^B_{rr}(r, \Omega) = \left. \frac{\partial^2 V^B(r, \Omega)}{\partial r^2} \right|_{r=r_{sat}(\Omega)} \quad \text{and} \quad V^{\sigma,B}_{rr}(r, \Omega) = \left. \frac{\partial^2 V^{\sigma,B}(r, \Omega)}{\partial r^2} \right|_{r=r_{sat}(\Omega)}. \tag{66}$$

Taking the radial second derivative of Eq. (23) at $r \geq r_t(\Omega)$ for V^B_{rr}

$$\frac{\partial^2 V^B(r, \Omega)}{\partial r^2} = 8\pi G \varrho_0 \frac{1}{r^3} \left[R^2 H(\Omega) + R H^2(\Omega) + \frac{1}{3} H^3(\Omega) \right], \tag{67}$$

the radial second derivative of Eq. (31) at $r > R$ for $V^{\sigma,B}_{rr}$

$$\frac{\partial^2 V^{\sigma,B}(r, \Omega)}{\partial r^2} = 8\pi G \sigma(\Omega) \frac{R^2}{r^3}, \tag{68}$$

and using Eqs. (33)–(34) for $\sigma(\Omega)$, we obtain $\delta E^B(r, \Omega) = 0$. Consequently, as for the direct topographical effect on gravity δA, the direct topographical effect on gradiometry δE only consists of the terrain roughness contribution δE^R that is

$$\delta E^R(r,\Omega) = V^R_{rr}(r,\Omega) - V^{\sigma,R}_{rr}(r,\Omega), \tag{69}$$

where

$$V^R_{rr}(r,\Omega) = \left.\frac{\partial^2 V^R(r,\Omega)}{\partial r^2}\right|_{r=r_{sat}(\Omega)} \quad \text{and} \quad V^{\sigma,R}_{rr}(r,\Omega) = \left.\frac{\partial^2 V^{\sigma,R}(r,\Omega)}{\partial r^2}\right|_{r=r_{sat}(\Omega)}. \tag{70}$$

Therefore, differentiating V^R (Eq. (25)) and $V^{\sigma,R}$ (Eq. (35)) twice with respect to r for A^R and $A^{\sigma,R}$, respectively, we obtain the expression for the direct topographical effect on gradiometry

$$\begin{aligned}\delta E(r,\Omega) = G\varrho_0 \int_{\Omega_0} \Bigg[& \left.\frac{\partial^2 \tilde{L}^{-1}(r,\psi,r')}{\partial r^2}\right|_{r'=R}^{r_t(\Omega')} - \left.\frac{\partial^2 \tilde{L}^{-1}(r,\psi,r')}{\partial r^2}\right|_{r'=R}^{r_t(\Omega)} \\ & -R^2[\tau(\Omega') - \tau(\Omega)] \frac{\partial^2 L^{-1}(r,\psi,R)}{\partial r^2}\Bigg]_{r=r_{sat}(\Omega)} d\Omega'.\end{aligned} \tag{71}$$

Here H_{sat} is the flying altitude of the satellite which takes the gradiometry measurements. For instance, the satellite GOCE with $H_{sat}(\Omega) = 250$ km.

2.5 Computations of the topographical effects

2.5.1 The integral Newton kernels for numerical computation

The distance $L(r,\psi,r')$, defined Eq. (18), can be written with the form $\sqrt{X}$, where X is the rational function

$$X = ax^2 + bx + c, \text{where} \quad x = 1, \;\; a = 1, \;\; b = -2r\cos\psi, \;\; c = r^2. \tag{72}$$

Thus, we can write the Newton kernel $\tilde{L}^{-1}(r,\psi,r')$ Eq. (19) with the form

$$\int \frac{x^2}{\sqrt{X}} dx. \tag{73}$$

To solve this integral, we use the following equations Gradshteyn and Ryzhik [12]

$$\int \frac{x^2 dx}{\sqrt{X}} = \left(\frac{x}{2a} - \frac{3b}{4a^2}\right)\sqrt{X} + \frac{3b^2 - 4ac}{8a^2}\int \frac{dx}{\sqrt{X}} \tag{74}$$

$$\int \frac{dx}{\sqrt{X}} = \frac{1}{\sqrt{a}} \ln|2\sqrt{aX} + 2ax + b| + C, \tag{75}$$

for $a > 0$ and where C is a constant value.

Replacing Eq. (75) and "our" notations in Eq. (74), we obtain the analytical expression:

$$\begin{aligned}\tilde{L}^{-1}(r,\psi,r') = \frac{1}{2}\Bigg[& (3r\cos\psi + r')\sqrt{r'^2 - 2rr'\cos\psi + r^2} + \\ & + r^2(3\cos^2\psi - 1)\ln\left|2\left(\sqrt{r'^2 - 2rr'\cos\psi + r^2} - r\cos\psi + r'\right)\right|\Bigg] + C\end{aligned} \tag{76}$$

Hence, the analytical expression $\tilde{L}_a^{-1}(r,\psi,r')$ from the list of integrals of Gradshteyn and Ryzhik [12]

$$\begin{aligned}\tilde{L}_a^{-1}(r,\psi,r') = &\frac{1}{2}[(3r\cos\psi + r')L(r,\psi,r')+ \\ &+r^2(3\cos^2\psi - 1)\ln|L(r,\psi,r') - r\cos\psi + r'|] + C_1(r,\psi),\end{aligned} \tag{77}$$

where $C_1(r,\psi)$ a “constant” which may depend only on the variables r and ψ. The computation of δN uses the expression of $\tilde{L}_a^{-1}(r,\psi,r')$.

2.5.2 First radial derivative of the Newton kernel

The radial derivative of the Newton kernel gives

$$\frac{\partial \tilde{L}^{-1}(r,\psi,r')}{\partial r} = \int_{r'} r'^2 \frac{\partial L^{-1}(r,\psi,r')}{\partial r} dr'. \tag{78}$$

By Eq. (18), since

$$(u^\alpha)' = \alpha u' u^{\alpha-1},$$

we readily get

$$\frac{\partial L^{-1}(r,\psi,r')}{\partial r} = -\frac{r - r'\cos\psi}{(r'^2 - 2rr'\cos\psi + r^2)^{3/2}}. \tag{79}$$

By substituting the expression (79) in (78), and integrating the equation, we obtain the analytical expression

$$\begin{aligned}\frac{\partial \tilde{L}_a^{-1}(r,\psi,r')}{\partial r} = &\left(3r^2\cos\psi + r(r' - 6\cos\psi^2 r') + \cos\psi r'^2\right)L^{-1}(r,\psi,r') \\ &+r(3\cos\psi^2 - 1)\ln|L(r,\psi,r') - r\cos\psi + r'| + C_2(r,\psi).\end{aligned} \tag{80}$$

The computation of δA uses the expression of $\partial\tilde{L}_a^{-1}(r,\psi,r')/\partial r$.

2.5.3 Second radial derivative of the Newton kernel

Expression of $\partial^2\tilde{L}_a^{-1}/\partial r^2$ emphasizes the second radial derivative of the Newton kernel, which has to be derived two times.

$$\frac{\partial^2 \tilde{L}^{-1}(r,\psi,r')}{\partial r^2} = \int_{r'} r'^2 \left(\frac{\partial^2 L^{-1}(r,\psi,r')}{\partial r^2}\right) dr' \tag{81}$$

where

$$\frac{\partial^2 L^{-1}(r,\psi,r')}{\partial r^2} = \frac{\partial}{\partial r}\left(\frac{\partial L^{-1}(r,\psi,r')}{\partial r}\right) \tag{82}$$

To obtain the second derivative of $L^{-1}(r,\psi,r')$, we derive over r, its first derivative, Eq. (79), using the property

$$\left(\frac{u}{v}\right)=\frac{u'v-uv'}{v^2}. \tag{83}$$

We finally obtain:

$$\frac{\partial}{\partial r}\left(\frac{\partial L^{-1}(r,\psi,r')}{\partial r}\right)=-\frac{1}{\left(r'^2-2rr'\cos\psi+r^2\right)^{3/2}}+\frac{3(r-r'\cos\psi)^2}{\left(r'^2-2rr'\cos\psi+r^2\right)^{5/2}} \tag{84}$$

Including Eq. (84) in Eq. (81), the second derivative of $\tilde{L}^{-1}(r,\psi,r')$ gives the followed integral:

$$\frac{\partial^2\tilde{L}^{-1}(r,\psi,r')}{\partial r^2}=\int_{r'}-\frac{r'^2}{\left(r'^2-2rr'\cos\psi+r^2\right)^{3/2}}+\frac{3r'^2(r-r'\cos\psi)^2}{\left(r'^2-2rr'\cos\psi+r^2\right)^{5/2}}dr' \tag{85}$$

After deriving a second time the reciprocal distance L^{-1} and integrating this equation, we obtain the analytical expression

$$\begin{aligned}\frac{\partial^2\tilde{L}_a^{-1}(r,\psi,r')}{\partial r^2}&=\left[3r^3\cos\psi+r^2\left(1-12\cos^2\psi\right)r'\right.\\&\left.+2r\cos\psi\left(6\cos^2\psi+1\right)r'^2+2\left(1-4\cos^2\psi\right)r'^3\right]L^{-3}(r,\psi,r')\\&+\left(3\cos^2\psi-1\right)\ln|L(r,\psi,r')-r\cos\psi+r'|+C_3(r,\psi),\end{aligned} \tag{86}$$

where

$$L^{-3}(r,\psi,r')=\left(r'^2-2rr'\cos\psi+r^2\right)^{-3/2}. \tag{87}$$

The computation of δE uses the expression of $\partial^2\tilde{L}_a^{-1}(r,\psi,r')/\partial r^2$.

3. Numerical studies

The determination of topographical effects from DEMs is a very time-consuming process, particularly when computations are required for large areas, such as a country or a continent. With a fine grid resolution, for instance, 50 m Quadratic Grid Resolutions (QGR) a unified spatial data structure, computations are beyond what a multi-processor computer can accomplish within a reasonable time-frame for an area such as Ireland. Numerical investigations have resulted that increasing the spatial resolution of DEM by a factor of two increases CPU computational time by a factor of fourteen.

Although the computational time is a factor to be taken into account, it is less important since it is the spatial resolution of DEM which is critical for improving accuracy required when a precise geoid is determined.

The Fast Fourier Transform (FFT), which relies on linearization and series expansions of the non-linear terrain effect integrals, provides a reduction in computational time by several orders of magnitude, compared to space domain integration methods

[13]. In the FFT, the higher-order terms of the radially integrated Newton kernel expressed by the Taylor series expansion are neglected. In addition, the reciprocal distance $1/L$ is approximated by the planar distance $1/\ell_0$ i.e. $\ell_0 = L(R, \psi, R)$ [14], Eq. (15)], which is a sufficiently good only if $\ell_0 > H(\Omega')$. This condition is violated for computing topographical effects in rough terrain.

3.1 The study area

Topographical effects in high elevated topography in Ireland are computed from 50 m QGR, Ireland with a relatively flat terrain surrounded by the ocean. The study area is 1° × 0.7° limited between longitudes −5.8° and −6.8° and latitudes 52.7° and 53.3°. A map of topographic heights of the study areas is presented in **Figure 8a**, and the distribution of topography in meters, counted in the percentage of the number of points in 50 m QGR is demonstrated in histograms **Figure 8b**.

3.2 The bound of integration area

The gravitational potential of Topographic Masses of finite thicknesses behaves like the potential of a thin layer when it is observed from a larger distance. This is explained by the behavior of integration kernels generating the potential of the gravitational potential $V^t(r, \Omega)$ and $V^c(r, \Omega)$. **Figure 9** illustrates the behavior of the two kernels for the determination of PITE and DTE relative to maximum elevation in Ireland, which is the *Carrauntoohil* elevation of 1039 m [15]. In the determination of the primary indirect topographical effect, when the angular distance Ψ between the computation point and integration point increases, the integration kernel,

$$K^c_{pite}(R, \psi, R) = R^2 \left(\frac{\tau(\Omega')}{L(R, \psi, R)} \right) \tag{88}$$

generating the potential of Helmert's condensation layer $V^c(r, \Omega)$ approaches the integration kernel

$$K^t_{pite}(R, \psi, H(\Omega')) = \tilde{L}^{-1}(R, \psi, r')\Big|_{r'=R}^{R+H(\Omega')} \tag{89}$$

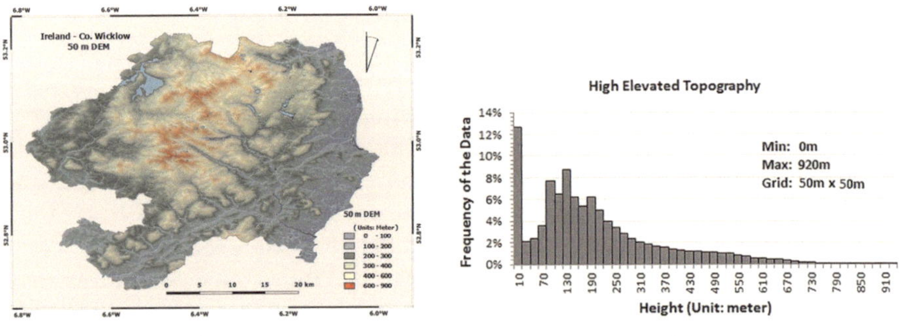

(a) (b)

Figure 8.
*Topographic heights over Co-Wicklow Ireland **Figure 8a** and distribution of elevations in meter, counted in percentage of the number of points from 50 m² DEM. (a) Topography - County Wicklow Ireland. (b) Distribution of topography.*

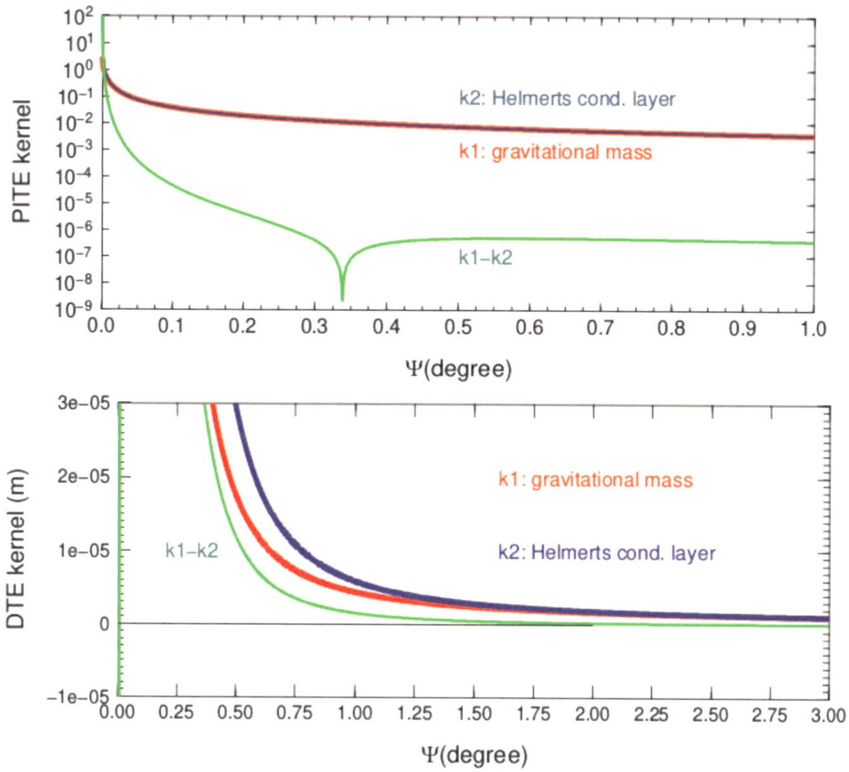

Figure 9.
The behavior of K_t and K_c in the dependence of varying the angular distance ψ in the determination of PITE and DTE in Ireland. The elevation of computation and integration point are fixed to 1039 m and 1 m respectively.

generating the gravitational potential $V^t(R,\Omega)$. Thus, the differences between two kernels $K^t_{pite} - K^c_{pite} = \delta K_{pite}$ decrease (see **Figure 9**-PITE kernel). In determination of DTE Eq. (57) a similar decrease occurs for the difference between two kernels $\left(K^t_{dte} - K^c_{dte} = \delta K_{dte}\right)$, when the integration kernel

$$K^c_{dte}(H(\Omega),\psi,R) = R^2\tau(\Omega)\frac{\partial L^{-1}(r,\psi,R)}{\partial r}, \tag{90}$$

generating the potential of Helmert's condensation layer $V^c(r,\Omega)$ approaches the integration kernel

$$K^t_{dte}(H(\Omega),\psi,H(\Omega')) = \frac{\partial \tilde{L}^{-1}(r,\psi,r')}{\partial r}\Bigg|_{r'=R}^{R+H(\Omega')}, \tag{91}$$

generating the gravitational potential $V^t(r,\Omega)$ of the TMs (see **Figure 9**-DTE). This also means that the magnitudes of δK_{pite} or δK_{dte} are largest in the immediate neighborhood of the computation point.

The choice of varying angular distance and fixing the elevation of integration point to 1039 m and computation point to 1 m or vice versa, enables us to determine the most attainable differences (maximum or minimum) between the kernels in question.

Furthermore, a numerical examination of δK_{pite} and δK_{dte}, controlled by varying topographic height $H(\Omega')$ at a fixed angular distance $\psi = 0.0001°$, in the immediate neighborhood of the computation point shows that the larger the height of the integration point, the larger the difference between these kernels, and therefore the topographical effects are stronger (see **Figure 10**).

Eqs. (36) and (57) assume the compensation is strictly local, which means $\delta A(\Omega)$ consists of the terrain roughness contribution only. The longitudinal profiles (see **Figure 11**) show that the Bouguer components of PITE have a larger contribution to topographical effects than the terrain roughness components. Since DTE does not contain a Bouguer component, the correlation between DTE and DEM is in general smaller than that for PITE.

3.3 Direct topographical effect on gravity

3.3.1 On the Earth's surface - residual effect δA_{surf}

The computation of the residual direct topographical effect on gravity on the Earth's surface δA_{surf} uses Eq. (57). **Table 1** shows the minimum, mean, maximum and root mean square (rms) of δA_{surf} in mGal.[5]

Due to the low elevation of topography in the test area (maximum 920 m), δA_{surf} is small. Numerical investigation with different grid resolutions shows that the flatter and lower the terrain, the smaller the reduction of the surface gravity observations in the remove step of the RCR procedure is. And a sparse griding reduces values of δA, so using as finer griding as possible is recommended.

Since the residual effect δA is rather small, we compared it with the total effect to estimate the efficiency of reduction by the Helmert condensation using the following expression

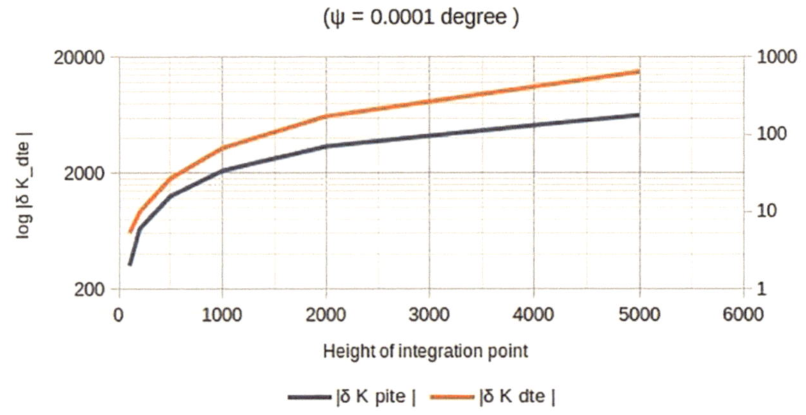

Figure 10.
Magnitudes of $|\delta K_{pite}|$ and $|\delta K_{dte}|$ in an immediate neighborhood ($\psi = 0.0001°$) of the computation point with $H(\Omega) = 1$ and varying height of integration point.

[5] Unit of gravity: 1 mGal = 10^{-5}m.s^{-2}.

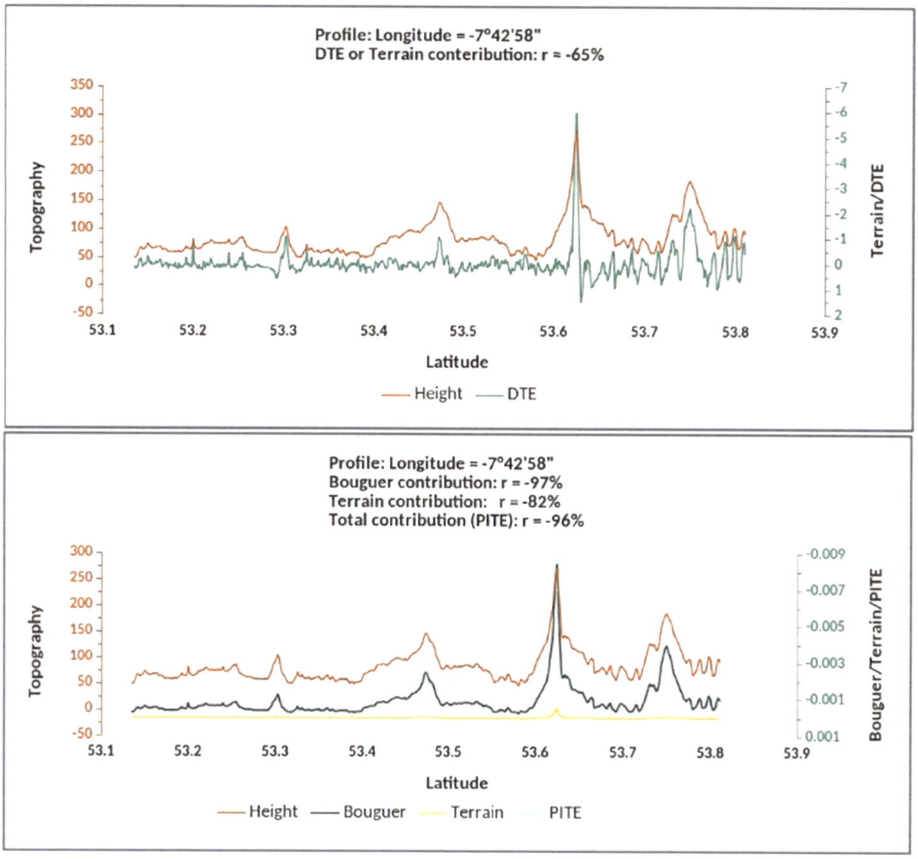

Figure 11.
Correlation between topographical effects and elevation of topography from 50m² QGRs.

$$A^t(r,\Omega) = G\varrho_0 \int_{\Omega_0} \frac{\partial \tilde{L}^{-1}(r,\psi,r')}{\partial r}\Bigg|_{r'=R}^{r_t(\Omega')}, \tag{92}$$

where $r = r_t(\Omega)$ (see **Table 1**). The total direct topographical effect on gravity A^t_{surf} which is the case when the Helmert condensation is not employed, is larger in amplitude than the residual effect δA_{surf}, and mostly negative. The values of A^t_{surf} are prevailingly distributed around 0 mGal in flat areas and are highly correlated with the topography.

	Min	Mean	Max	rms
Topographic Heights (m)	0	58.600	920.000	±200.000
δA_{surf} (mGal)	−17.4	0.006	5.7	±0.7
δA^t_{surf} (mGal)	−27.5	−0.6	2.6	±1.5

Table 1.
Direct topographical effect δA on gravity.

3.3.2 At satellite altitudes—residual effect δA_{sat}

The computation of the residual direct topographical effect on satellite gravity δA_{sat} uses Eq. (59) for $r_{sat} = r_{GRACE} = R + 400$ km..

The residual effect on gravity δA_{sat} computed at satellite altitude is of the order of 10^{-5} mGal, that is significantly smaller in comparison with δA_{surf} which is of the order of 10's to 100's of mGals (see **Table 2**). The total direct topographical effect on satellite gravity A^t_{sat} also uses Eq. (92) for $r = r_{GRACE}$. Comparing the residual effect δA_{sat} with the total effect A^t_{sat}, we see that the total effect A^t_{sat} has a long-wavelength feature. This shows that the gravitational signal of the topography is attenuated when going to satellite altitudes in such a way that short wavelengths of the gravitation are attenuated faster than long wavelengths. The fact that the total effect A^t_{sat} is of the order of 10's of mGals and the residual effect δA_{sat} almost vanishes shows that the compensation of the masses by the Helmert condensation is an efficient way to process satellite gravity data. However, this is not the case for surface gravity data. Note that a coarser griding does not affect the results of A^t_{sat}.

3.4 Direct topographical effect on gradiometry—residual effect δE

The computation for the direct topographical effect on gradiometry δE uses Eq. (71) for $r_{sat} = r_{GOCE} = R + 250$ km. **Table 3** shows the minimum, mean, maximum and rms values of δE in Eötvös.[6] The values of δE given in **Table 3** are of the order of 0.1 m $\overset{\circ}{E}$. The coarser topographical griding does not affect the results of the topographical effect δE on gradiometry. Even though δE has tiny values, these are still slightly correlated with the topography. As the measuring accuracy of the second derivatives by the GOCE gradiometer is 10 m$\overset{\circ}{E}$, the effect of δE should still be taken into account in the RCR procedure, especially in mountainous area.

	Min	Mean	Max	rms
δA_{sat} (10^{-4} mGal)	−3.1	−0.002	138.9	±6.7
δA^t_{sat} (mGal)	−0.87	−0.74	−0.51	±0.8

Table 2.
Direct topographical effect δA on satellite gravity.

	Min	Mean	Max	rms
$\delta E \left(m\overset{\circ}{E}\right)$	−2.13	−0.00001	0.7	±0.10
$V^t_{rr} \left(\overset{\circ}{E}\right)$	0.01	0.7	0.11	±0.02

Table 3.
Direct topographical effect δE on gradiometry.

[6] unit of gradiometry: 1∘E = 10^{-9}m/s^2/m.

The total direct topographical effect V_{rr}^t on gradiometry is computed by

$$V_{rr}^t(r,\Omega) = G\varrho_0 \int_{\Omega_0} \frac{\partial^2 \tilde{L}^{-1}(r,\psi,r')}{\partial r^2} \Bigg|_{r'=R}^{r_t(\Omega')}, \tag{93}$$

where $r = r_{GOCE}$. **Table 3** shows that the values of V_{rr}^t are two orders in magnitude larger than δE. The fact that the total effect V_{rr}^t is much larger in amplitude than the residual effect δE shows that the compensation of the masses by the Helmert condensation is again an efficient way to process satellite gravity gradient data as well as the satellite gravity data. The distribution of the total effect V_{rr}^t between the maximum and minimum values is much more homogeneous compared to the distribution of the residual effect δE with most values around 0$\overset{\circ}{\text{E}}$.

3.5 Topographical effect on geoid heights

3.5.1 The effect of PITE on geoid heights

The primary indirect topographical effects on the geoid height determination are computed by Eq. (42),

$$\delta N(R,\Omega) = \frac{\delta V(R,\Omega)}{\gamma_Q}.$$

Figure 12b illustrates the effect of PITE in a high elevated topography at 50 m grid resolution.

3.5.2 The effect of DTE on geoid heights

The direct topographical effects on geoid heights, δD, are computed by applying Stokes' integral to DTE as a known function $f(\Omega)$ distributed in the near-zone spherical cap C_{ψ_0}

$$\delta D^{\ell,\psi_0}(\Omega) = \frac{R}{4\pi} \int_{C_{\psi_0}} f^{\ell}(\Omega') S^{\ell}(\psi) d\Omega', \tag{94}$$

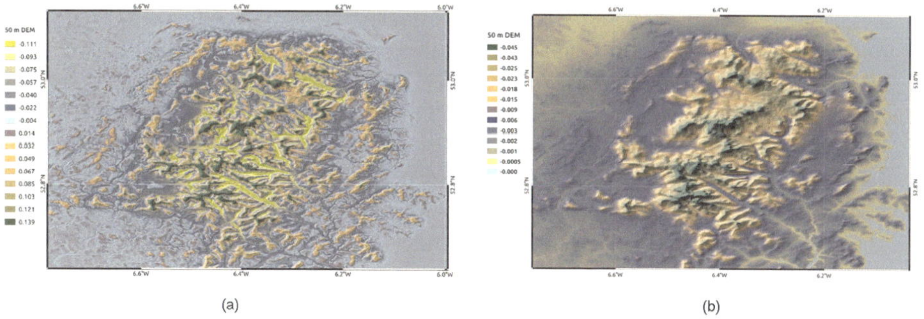

Figure 12.
Illustration of DTE and PITE on geoid undulation for High elevated topography in Ireland at 50 meter grid resolutions. (a) Effects of DTE on geoid undulation. (b) Effects of PITE on geoid undulation.

where $\delta D^{\ell,\psi_0}$ is a higher-degree (presented as superscript ℓ) correction to the geoid height, C_{ψ_0} is a spherical cap of radius ψ_0 (for this study $\psi_0 = 1\circ$ is chosen) and $S^{\ell}(\psi)$ is the spheroidal Stokes function, [16]. The effect of DTE on geoid heights is also presented in **Figure 12a**.

4. Conclusions

The RCR procedure for gravity and gravimetry is summarized in **Figure 2**. It shows the motivation for correcting the data by the effects of the topographical masses. These are the direct or indirect topographical effects computed at surface or satellite altitude. This study summarized the field corrections for the determinations of geoids by terrestrial data or the latest satellite missions. The integral of newton presented in Sec:2.1.1 in the form of the gravitational potential of topography Eq. (16), proves necessary to the expression of topographical effects. It made it possible to understand the observations made either $\partial V/\partial r$ or second derivatives $\partial^2 V/\partial r^2$. Furthermore, make it possible to fulfil the conditions necessary to determine a model of geoids whose centimetre precision is required.

Presentation of the *RCR* method allowed us to understand the interest in determining the effects of topographic masses in order to correct gravimetric and gradiometric measurements and to be able to apply the calculations to determine a geoid model.

In order to express the effects, we need to know the topographic masses defined by their height and density. We have seen that Newton's integral for determining the geoid allows expressing the effects. The topography effects on the satellite measurements are calculated in a way similar to the effects on ground measurements.

The effects of topographic masses have much less impact on gravity measurements at the satellite level than on ground measurements.

Computing topographical effects for large areas is a very time-consuming process. Increasing the resolution of sampled DEM by a factor of 2 (e.g. from 100 m to 50 m quadrangle) increases the number of data by a factor of 4, and it increases the computational time by a factor of approximately 14. Thus, it is suggested to restrict the integration area to a small area of radius ψ_0 around the computation point.

A sparse grid size, particularly in rugged areas, is not sufficient to express the irregularities of the terrain and thus does not reveal properly the contribution to geoidal height due to terrain height variations.

With a tiny grid step size, the magnitude of the Bouguer component becomes comparable with that of the terrain roughness component, which reduces the correlation between DTE and PITE. Since DTE does not contain a Bouguer component, the correlation between DTE and DEM is generally smaller than that for PITE.

Numerical investigation shows that the Bouguer components of PITE have a larger contribution to topographical effects than the terrain roughness components.

Numerical examination of Kernel's, controlled by varying topographic height at a fixed angular distance, in the immediate neighborhood of the computation point, shows that the larger the height of the integration point, the more significant the difference between these kernels, and therefore, the topographical effects are more substantial.

Comparing the results with different grid sizes shows (not shown here) an improvement in computation accuracy. Contrary to our expectations, it is not the case for calculations at satellite altitudes, so griding can be reduced, and a more refined grid does not change a long-wavelength feature of V^t_{rr}.

DOI: http://dx.doi.org/10.5772/intechopen.109894

Acknowledgements

The authors would like to thank Dr. Pat Gill (TUS) and Dr. Nic Wilson (UCC) for their comments that have helped to improve the manuscript.

Author details

Sajjad Sajjadi[1]* and Zdenek Martinec[2]

1 Technological University of the Shannon (TUS), Ireland

2 Dublin Institute for Advanced Studies Geophysics Section, Ireland

*Address all correspondence to: sajjad.sajjadi@tus.ie

References

[1] Sajjadi S, Martinec Z, Prendergast P, Hagedoorn J, Šachl L. The stability criterion for downward continuation of surface gravity data with various spatial resolutions over Ireland. Studia Geophysica et Geodaetica. 2021;**65**(3): 219-234

[2] Hofmann-Wellenhof B, Moritz H. Physical Geodesy. 2nd ed. Heidelberg: Springer-Verlag; 2006. p. 404

[3] Martinec Z. Green's function solution to spherical gradiometric boundary-value problems. Journal of Geodesy. 2003;**77**(1):41-49

[4] Heiskanen WA, Moritz H. Physical Geodesy (Book on Physical Geodesy Covering Potential Theory, Gravity Fields, Gravimetric and Astrogeodetic Methods, Statistical Analysis, etc). Harvard. 1967

[5] Forsberg R. A Study of Terrain Reductions, Density Anomalies and Geophysical Inversion Methods in Gravity Field Modelling. Harvard: Ohio State Univ Columbus Dept Of Geodetic Science and Surveying; 1984

[6] Heck B. On Helmert's methods of condensation. Journal of Geodesy. 2003; **77**(3):155-170

[7] Martinec Z. The indirect effect of topography in the Stokes-Helmert technique for a spherical approximation of the geoid. Manuscripta Geodaetica. 1994;**19**:213-219

[8] Martinec Z, Vanicek P. Direct topographical effect of Helmert's condensation for a spherical approximation of the geoid. Manuscripta Geodaetica. 1994;**19**(5):257-257

[9] Wichiencharoen C. The Indirect Effects on the Computation of Geoids Undulations. Columbus: Department of Geodetic Science, The Ohio State University; 1982

[10] Martinec Z. Boundary-Value Problems for Gravimetric Determination of a Precise Geoid. Heidelberg: Springer-Verlag; 1998. Lecture Notes in Earth Sciences. p. 223

[11] Helmert FR. Die matematischen und physikalischen Theorien der höheren Geodäsie. Vol. 2. Leipzig (reprinted in 1962 by Minerva GMBH, Frankfurt/ Main): B.G. Teubner; 1884. p. 572

[12] Gradshteyn IS, Ryzhik M. Table of Integrals, Series and Products. 7th ed. Orlando, Florida: Academic Press; 2007. 1171 p

[13] Forsberg R. Gravity field terrain effect computations by fft. Bulletin géodésique. 1985;**59**(4):342-360

[14] OCD Omang and René Forsberg. How to handle topography in practical geoid determination: Three examples. Journal of Geodesy. 2000;**74**(6):458-466

[15] OSi. Discovery series, mapsheet 78 1: 50,000. 1995

[16] Vanicek P. The canadian geoid-stokesian approach. Manuscripta Geodaetic. 1987;**12**:86-98

Chapter 6

Artificial Intelligence Techniques for Observation of Earth's Changes

Eman A. Alshari and Bharti W. Gawali

Abstract

This chapter discusses the primary components that contribute to the observation of Earth's changes, including Land Observation Satellites, land classification techniques and their stages of development, and Machine Learning Techniques. It will give a comprehensive summary of the development stages of high-resolution satellites. It also details land classification with artificial intelligence algorithms. It will also give knowledge of classification methodologies from various Fundamentals of Machine Learning Classifiers: Pixel-based (PB), Sub-pixel-based (SPB), Object-based (OB), Knowledge-based (KB), Rule-based (RB), Distance-based (DB), Neural-based (NB), Parameter Based (PB), object-based image analysis (OBIA). It includes several different classifiers for LULC Classification. This chapter will include two applications for land observation satellites: The first is land use and land cover change observation with a practical example (study land use and land cover classification for Sana'a of Yemen as a case study from 1980 to 2020). The second application is satellite altimetry monitoring changes in mean sea level. The most significant contributions of it are the integration of these components. This chapter will be crucial in helping future researchers comprehend this topic. It will aid them in selecting the most appropriate and effective satellites for monitoring Earth's changes and the most efficient classifier for their research.

Keywords: earth changes observation (ECO), machine learning (ML), high-resolution satellites, artificial intelligence (AI), land use land cover (LULC), land observation satellites (LOS)

1. Introduction

The art and science of measuring the planet earth through sensors or satellites are known as remote sensing, which, together with GIS technology, become an essential aid in collecting data about the Earth. The overall purpose of image collection is to naturally classify all pixels in an image into land cover groupings or subjects. In LULC categorization, their unique Artificial Intelligence approaches play a key role. Therefore, the work is interesting for this book. The work represents an overview of the application of artificial intelligence in detecting Land Use Land Cover (LULC) changes on Earth. This chapter reviews Earth observation satellites and their development, brings synthetic intelligence procedures for Land Use Land Cover (supervised and unsupervised methods) and fundamentals of ML classifiers, and ends with the challenges of AI techniques for LULC classification and conclusion [1].

The land has become a gigantic and immense resource of economic that cannot be underestimated in any region, where Earth Changes Observation led to serving the country's economic, political, and social needs. Understanding land changes is vital for land resource management and assessing the technology's potential, where LULC change detection assists policymakers in understanding the dynamics of environmental change to ensure long-term growth. As a result, LULC feature identification has become an essential research topic, necessitating the development of a robust and reliable LULC classification [2].

Land use cover is necessary to make up the land's physically present and visible surface components [3] which allow researchers to investigate landscape patterns and features, which are essential to understanding land size, location, and condition of the size, structure, and state of the ecosystem [4]. The importance of land classification stems from using a specific piece of land that may be linked to significant price differences, necessitating a well-defined land categorization. The price (development) of land underneath houses, for example, may be drastically different from the price (impact) of agricultural land [5]. As a result, remote sensing and geographic information systems (GIS) have become essential aid in collecting data about the Earth, and It's considered critical for risk assessment and monitoring land degradation and conservation [6, 7].

It is possible to collect data across large geographical areas and define natural qualities or physical items on the ground. Analyzed surface areas and objects regularly, tracked their changes over time and combined this data [8] by several decision support systems relying on remote sensing for land use and land cover (LULC) detection [9]. As considered, Artificial intelligence (AI) is a technique solid and active in studying and developing computers or computational systems that can accomplish tasks that would need human intelligence in this field. According to the innovative operation, artificial intelligence algorithms play an essential part in LULC, where these classifiers for LULC classification can be split into different classes. Arrangement strategies are divided into two categories. The first category is traditional machine learning (complex classification) (unsupervised and semi-supervised, supervised), contemporary machine learning, and based knowledge discovery are examples of challenging classes. The second category is soft classification [10].

AI is useful because it enables software to execute human-like functions like reasoning, planning, communication, and perception more effectively, efficiently, and at a lower cost [11]. Quantum computing has much potential for improving AI and machine learning algorithms. Although the technology is currently out of reach for most people, Microsoft, Amazon, and IBM are making quantum computing resources and simulations available via cloud models [12].

2. Satellites for land observation

Earth changes observation is a ground photography scanning system that collects, stores, analyses, and displays land photographs using remote sensors at regular intervals [13]. Earth changes observation is used to detect changes in land cover over time and monitor and analyses changes in the natural and built environments through the land's physical, computational, and biological systems, among other things, according to the Land Observation Group [7].

Land observation satellites are satellites designed to view the Earth from orbit and are used for various reasons, including mapping, environmental monitoring, meteorology, and other applications. They typically include remote sensors and wireless devices. Ground imaging satellites, which capture photos from satellites, are the most frequent type [8].

The spread of satellite launches in most technologically advanced countries has led to a new shift, in general, exploring land uses, land cover, and earth changes observation [14]. So the increasing need to continue developing remote sensing satellites to monitor the ground and know the current limitations to launch a revolution in space and technological development requires a careful study of the capabilities and challenges. This study will provide a comprehensive survey in this field to find valuable details [15].

2.1 Types of land observation satellites

The resolution of an image refers to the potential detail provided by the imagery. Resolution refers to the smallest size an object or element can be represented in an image. Higher resolution means that pixel sizes are smaller, which provides more details. Then **Figures 1** and **2** show the types of resolution of Land Observation Satellites (LOS) with some significant features [15].

2.2 Development of land observation satellites (LOS)

The spatial resolution was the fundamental distinction between high-resolution and low- and medium-resolution satellites. The higher spatial resolution (0.5–1.5 m/pxl) elevates the image qualities - from an unexpectedly detailed image due to the length units to a ratio of pixel values that provides the user with greater precision. It is also surrounded by the latest Optical technologies onboard the high-resolution satellites for remote sensing in addition to a high visit (≤ one day), which allows for observing the current conditions of the Earth's surface [16]. Another great

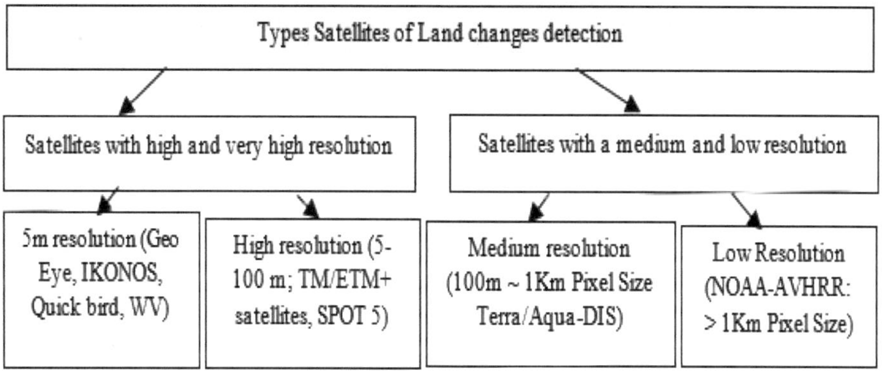

Figure 1.
Types satellites of land changes detection.

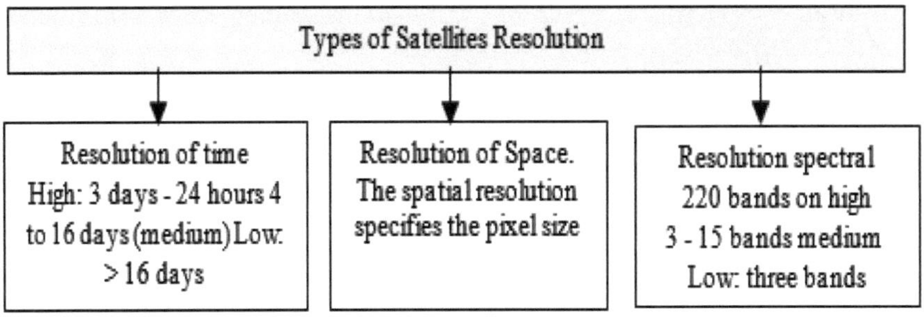

Figure 2.
Types of resolution satellites.

feature of high-definition satellites for commercial data is the assignment of a dedicated mission in terms of the ability to commission a high-resolution satellite to take a new photo. Always the reason why HRSI has a unique characteristic is that it is clearly and publicly commercially based data (here applied to the saying: you get for what you pay [17]. The fact that high-resolution satellites are commercial offers them the following advantages:

- Filtering the image to 100% of the region covering your use case.
- Get free samples, and a sneak peek.
- Instant access to the pricing because it is determined automatically.
- Three working days to receive the image.
- When used offline, high-resolution photos can be processed directly in the browser, avoiding FTP and downloading high-resolution satellite images for further analysis. It saves time, effort, storage space, and the cost of specialized tools.

The literature describes the benefits of high-resolution satellites, quick delivery, and fine details [18]. After the appearance of high-resolution satellites was removed, this section presents an overview of the high-resolution expected. The following are details on the development of (LOS) in **Tables 1** and **2**: (1) from 1999 to 2010, the first generation. (2) 2010–2015, the second generation. (3) 2015–2020 third generation (4) The Future—Fourth Generation 2020 As stated in the references [19], the data was gathered from a variety of credible and recent sources. The investigations indicated that new and exciting breakthroughs would arise. Many satellites with high spatial accuracy will occur in the next century's early years. The inaccuracy of high-resolution spatial image data is better, but it is also more expensive [20]. However, there is still a demand for medium-resolution satellites because of the benefits that commercial satellites have and lack [21]. They're also less priced and arrive faster. It was also mentioned that the launch of more remote sensing satellites with higher accuracy than currently available will have increased. **Tables 1** and **2** [22] provide details on high-resolution satellite launches.

Country	Spacecraft	Date	Resolution
USA	IKONOS	1999	1 m
	QuickBird-1	2000	0.8 m
	EROS-A1	2000	1.5 m
	Preview-3	2001	1 m
	Preview-4	2001	1 m
	EROS-A	2001	1 m
	Quick Bird	2001	0.61 m
	WorldView-1	2007	0.5 m
	WorldView-2	2009	0.5 m
	GeoEye-1	2008	0.46 m
INDIA	IRS-P5	2002	2.5 m
	IRS-P6	2001	
	IRS-2A	2003	1 m
	Cartosat-2	2007	0.8 m
	Cartosat-2A	2008	
	Cartosat-1	2005	2.5 m
Russia	Kometa-20	2000	2 m
Canada	Radarsat2	2002	3 m radar
China/Brasil	CBERS-3	2002	3 m
China/Brasil	CBERS-4	2002	3 m
Japan	ALOS	2002	2.5 m
France	SPOT-5	2002	5 m
Taiwan, land & ocean	ROCCAT-2	2002	
DLR, Radar	TerraSAR,	2004	1 m
Emirates	DubaiSat-1	2009	
Italy	CSK-1	2007	
Italy	CSK-2	2007	
Italy	CSK-3	2008	
South Korea	Kompakt-2	2006	
Country	Spacecraft	date	Resolution
Canada	RADARSAT-2	2007	
Spain	(SEOSat)	2007	
USA	WorldView-3	2014	3 0.4 m PAN
	SkySat —1	2013	0.8 m PAN and 1.0 m MS
	SkySat —2	2014	0.8 m PAN and 1.0 m MS
	Alat-2A	2010	2.5 m PAN and a 10 m MS
	Sentinel-1	2014	

Country	Spacecraft	Date	Resolution
	VNREDSat-1A	2013	2.5 m
CHINA	Gaofen,GF1	2013	
	Gaofen, GF2	2014	0.8 m and a MS 3.2 m.
	China-Brazil-Earth Resources-Satellite (CBERS)-4	2014	
	TripleSat-1 to 3	2015	
INDIA	Cartosat-2B	2010	
Algeria	AlSat-2A	2010	
Italy	CSK-4	2010	
Spain	Deimos-2	2014	
Emirates	Dubai's-2	2013	
Kazakhstan	(KazEOSat-1)	2014	
South Korea	Kompakt-3	2012	
South Korea	Kompakt-3A	2015	
Nigeria	Nigeria-2,	2011	
France	Pleiades-1A	2011	
France	Pleiades-1B	2012	
France, Azerbaijan	SPOT-6	2012	
France, Azerbaijan	SPOT-7	2014	
United Kingdom, China	TripleSat-1, —2,-3	2015	
Vietnam	VNREDSat-1A	2013	2.5 m PAN and 10 m MS
USA	WorldView-4	2016	31 cm PAN and 1.24 m MS
	SkySat —3	2016	0.8 m PAN and 1.0 m MS
	SkySat —4 to —7	2016	0.8 m PAN and 1.0 m MS
	SkySat −8 to −13	2017	0.8 m PAN and 1.0 m MS
EUROPEAN	Alat-2B	2016	2.5 m PAN and a 10 m MS
	Sentinel-1B	2016	
	Sentinel-2A	2015	
	Sentinel-2B	2017	
	Sentinel-3A	2016	
	Sentinel-3B	2018	
CHINA	Gaofen, GF4	2015	
	Gaofen, GF3	2016	
	Gaofen,GF5,GF6	2017	
	Gaofen, GF7	2018	
	Zhuhai	2018	

Country	Spacecraft	Date	Resolution
	SSTL-S1–4	2018	1 m PAN mode and m MS
INDIA	Cartosat-2C	2016	
	Cartosat-2D	2017	
	Cartosat-2E	2017	
	Cartosat-2F	2018	
	Cartosat-3	2019 (planned)	
	GEO Imaging Satellite (GISAT)	2019	42 to 318 m
	HRSAT 1A, 1B, and 1C	2020 (planned)	
	resources-3S	2019 (planned)	
	resources-3SA	2020 (planned)	
Japan	ALOS-3	2020 (planned)	
Algeria	Alat-2B	2016	
South Korea	CAS500–1 and — 2	2020 (planned)	
Italy	CSG-1	2019 (planned)	
Italy	CSG-2	2020 (planned)	
United Arab Emirates	Khalifasat	2018	
Morocco	Mohammed VI-A	2017	
Morocco	Mohammed VI-B	2018	
Argentina	The Aleph-1	2018	
Peru	Peru's-1	2016	
Canada	RADARSAT,	2019	
Kingdom, China	SSTL-S1	2018	
United Kingdom	Vivid-i 1 to 5	2019	
Venezuela	VRSS-2	2017	

MS = multispectral and PAN = panchromatic [2].

Table 1.
High-resolution optical space sensors for the first generation, 1999–2020.

Country	The Spacecraft	Date
USA	(1HOPSat)	2020 (planned)
EUROPEAN	(FLEX)	2022 (planned)
	Sentinel-1C	2021 (Planned)
	Sentinel-1D	2023 (Planned)
	Sentinel-2C	2020 (planned)
	Sentinel-2D	2021 (planned)
	Sentinel-3C	2020 (planned)
	Sentinel-3D	2022 (planned)

Country	The Spacecraft	Date
INDIA	NISAR	2021 (planned)
South Korea	Kompakt-7	2021 (planned)

Table 2.
High-resolution optical space sensors for the fourth generation, 2020-future [2].

3. AI approaches to analyze LULC

Recently, Classifiers that create exact LULC maps have been in high demand. Dependable Information is required from remotely sensed pictures, even on high-dimensional, complex data. Machine Learning Classifiers have a significant role in giving good classification results. Several aspects influence the accuracy of classified maps, including training sample size, training sample quality, thematic correctness, classifier choice, study region size, etc. Understanding these criteria will aid in achieving the highest classification accuracy feasible for a given need [23]. Big Data challenges arise when classification tasks involving multiple satellite photos and features become computationally intensive. To identify AI techniques for LULC of ML, as described in **Figure 3**, this section from this chapter searched deeply for several methods that supply a significant impetus for future readers to develop ML techniques. As stated in **Table 3**, we investigated the foundations of different AI classification algorithms in this chapter.

To provide new readers knowledge about different LULC system foundations as follows: Pixel-based (PB), Sub-pixel-based (SPB), Object-based(OB), Knowledge-based(KB), Rule-based(RB), Distance-based(DB), Neural-based(NB), Parameter Based(PB), object-based image analysis (OBIA). It includes several different classifiers for LULC that are as follows: (Random Forst -RF, Tree Decision Classification - TDC, Maximum Likelihood Classifier –MLC, Spectral Angle Mapper Classification - SAM, Support Vector Machine -SVM, K-Nearest Neighbor – KNN, Minimum Distance Classification - MDC, Artificial Neural Networks - ANN, Mahalanobis, Maximum Entropy, Parallelepiped, Boosting, Normal Bayes, ISOData, and K-means) [24].

3.1 Supervised methods

Classification Supervised (human-guided): This is based on the idea that a user can select sample pixels in an image representing different classes and then tell image

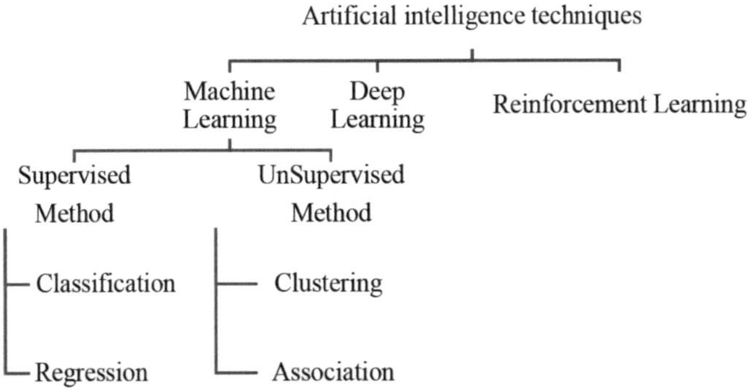

Figure 3.
Types of AI techniques for LULC classification.

	Supervised method	Unsupervised method
Process	In a supervised learning model, input and output variables will be given	In an unsupervised learning model, only input data will be given
Input data	Algorithms are trained using labeled data	Algorithms are used against data that is not labeled
Algorithms used	Support vector machine, neural network, linear and logistics regression, random forest, and classification trees	Unsupervised algorithms can be divided into different categories: cluster algorithms, K-means, hierarchical clustering, etc.
Computational complexity	Supervised learning is a more straightforward method	Unsupervised learning is computationally complex
Use of data	A supervised learning model uses training data to learn a link between input and output	Unsupervised learning does not use output data
Accuracy of results	Highly accurate and trustworthy method	Less accurate and trustworthy method
Real-time learning	The learning method takes place offline	The learning method takes place in real-time
Number of classes	The number of classes is known	The number of classes is not known
Main drawback	Classifying big data can be a real challenge in supervised learning	Cannot get precise information regarding data sorting, and the output as data used in unsupervised learning is labeled and unknown

Table 3.
Comparison of supervised and unsupervised methods.

processing software to use these training sites as references when classifying the rest of the pixels in the image [25]. The user's knowledge is used to choose training locations (testing sets or input classes). The user defines the boundaries for how similar they must be to group pixels together. These bounds are usually determined using the spectral properties of the training zone, plus or minus a defined increment (sometimes based on "brightness" or strength of reflection in specific spectral bands). In supervised learning [26], you use well-labeled data to train the algorithm.

It signifies that certain information has already been tagged with the appropriate response. It's analogous to learning with a teacher or supervisor present. A supervised learning system that learns from labeled training data can predict unexpected data outputs. It is possible to design, scale, and deploy accurate supervised machine learning. A group of highly skilled data scientists must devote time and technical skills to build a data science model. Data scientists must also keep their models up to date to ensure accurate insights even if the data changes [27]. The different supervised approaches are shown in **Figure 4** [27]: classification and regression.

3.1.1 Type of supervised methods

Regression: A method for predicting a single output value using training data, for example, uses regression to predict a property's price based on training data. Other input variables include location, dwelling size, and other aspects [28].

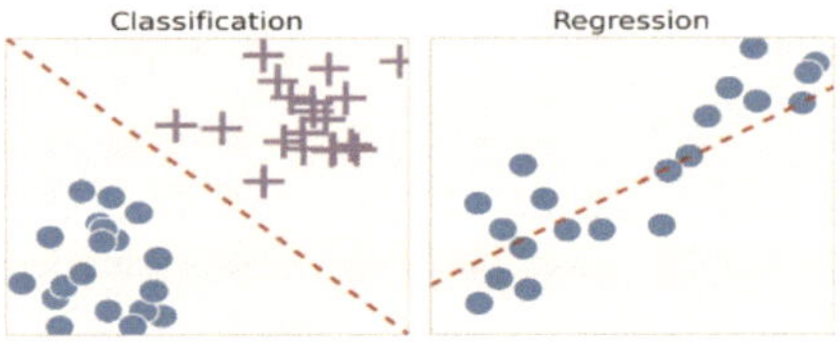

Figure 4.
Classification and regression from supervised techniques.

Classification: "classification" refers to categorizing output into different groupings. Binary type is when an algorithm splits data into two groups. The multiclass category is choosing between more than two classifications [29].

3.2 Unsupervised methods

The model is not supervised in unsupervised learning. Instead, allowing the model to determine what it requires would be beneficial. It mainly deals with data that has not been labeled. Unsupervised learning algorithms allow for more complex processing tasks than supervised learning algorithms. On the other hand, unsupervised learning is potentially more unpredictable than natural learning systems such as deep learning and reinforcement learning [30]. Classification Unsupervised (according to software): This classification is based on software analysis of an image without using user-supplied example classes. The computer uses algorithms to determine which pixels are connected and categorize them accordingly. **Figure 5** [31] shows this well.

3.2.1 Type of unsupervised method

Clustering When it comes to unsupervised learning, clustering is a crucial idea. Its primary purpose is discovering a structure or pattern in uncategorized data [31]. If natural clusters (groups) exist in the data, clustering algorithms will analyze and find them [32].

Associated: Associated: You can create associations between data elements in massive databases using association rules. This unsupervised method searches extensive databases for intriguing correlations between variables. For example, people who purchase a new home are more likely to buy new furnishings [33].

3.2.2 Comparison between supervised and unsupervised methods

There are several different Supervised and Unsupervised Methods methods, explained in **Table 3** down [34]:

Figure 5.
Types of unsupervised techniques.

3.3 The fundamentals of ML classifiers

3.3.1 Pixel-based classification

Pixel depiction approaches, such as the model remote recognizing image request technique, assume that each Pixel is pure and is commonly referred to as a single land spread type. Using this technique, distantly identifying imagery is perceived as a collection of pixels containing alarming data. As a result, extra standard components and their changes (for example, head sections, vegetation records, and so on) represent a pixel classifier commitment. Pixel-wise portrayal estimations can be divided into autonomous and supervised game plans. Using single classifiers, far distinct image is sorted into multiple classes based on the trademark groupings of the image without the usage of ready data or primary data on the review zone [35].

3.3.2 Sub-pixel-wise based

According to pixel-wise, far from identifying picture-gathering processes, each image pixel has only one land use kind. Regardless, such a notion is typically erroneous for medium and coarse objective imaging due to the unpredictability of scenes when viewed about the spatial target of a distant identifying picture. On the other side, using hard-plan plans decreases land use spread and helps representation accuracy. Because the areal degree of each land use may be precisely quantified, subpixel collection technologies are a popular alternative. Primary subpixel depiction backslides displaying backslide tree analysis, and supernatural mix assessment have been designed to address the mixing pixel issue. Overall, each Pixel receives fragmented enrolments with the soft depiction, and the contrasting areal degree of each class may be assessed [36].

3.3.3 Object-based classification

The spatial features of each Pixel as they relate to one another are considered while classifying a small collection of pixels. A pixel collection would be used as a preparation model for the classification algorithm. The classification algorithm would produce a class forecast for pixels. Object-based approaches divide images and route image requests to things rather than pixels, resulting in picture conflicts. In picture division approaches, mysterious, spatial, textural, and crucial information highlight picture objects. These articles have also mastered using unnatural and other critical models. Object-based techniques have greatly improved accuracy in multiple investigations [37] because different picture pixels make up a geographic item.

3.3.4 Image segmentation and object-based image analysis (OBIA)

A high-level image is broken down into numerous homogeneous components, each distinct. Image and article-based picture divisions, utilized in content-based image recovery, clinical imaging, object revelation, and other areas, are the third and fourth critical social events. Furthermore, Kitting and Land grebe, who later invented the ECHO classifier, made an early picture division application in the distance differentiating sector. Spatial important Information has been added to the calculations to separate far-flung identifying images, such as region construction, Markovian processes Jackson and Lange], watershed systems, and various evened-out computations.

A region is generated using the region creation approach by sharply isolating each neighboring Pixel's features from the space's mean. The pixels with tiny differences are distributed throughout the area. As a result, while each zone contains spatially coterminous and homogeneous pixels, there is evident variability at distinct locations [38].

3.3.5 Knowledge-based

In terms of technology, the branch of artificial intelligence has improved in recent years. Rule-based, data-driven, ensemble and reinforcement learning methods are mentioned. The divisions and algorithms for this classification type are listed in this section. In recent years, artificial intelligence has changed the way individuals think about new ideas. Rule-based methodology, data-driven methodologies, ensemble techniques, and reinforcement learning approaches are discussed. This file [39] contains all of the order's divisions and calculations.

3.3.6 Rule-based methodologies

Rule-based methods, which were the focus of early AI research, extract crucial Information from vast data using master data, agreed-upon rules, and reasoning techniques. The methodology is incredibly logical (predictable from the interaction of rational human thinking) but lacks adaptability, making it analogous to someone born with predetermined knowledge. The criteria cannot be updated once the model has been established, leaving the user powerless to solve new difficulties for which no standards have been developed. The most commonly used principle-based systems for information disclosure from remote detecting data are master frameworks, decision trees, and affiliation rule learning [40].

3.3.7 Distance-based

The least distance classifier (MDC) is an excellent design based on the distance between pixels in the highlight space. It is commonly thought that highlight focuses of the same class are grouped in component space. The mean vector governed by this element focuses as the class's focal point, and the covariance network represents the scattering of encompassing focuses. Every type has its own set of estimates for focus. The fundamental postulate of the similitude measure is that if the highlight contrasts between two modes are below a given edge, the two modes are believed to be comparable. It addresses a variety of dynamic districts by utilizing a region encompassed by various preparation test centers and calculating test similitude using distance as the significant criterion [37].

3.3.8 Neural based

During the learning process, a neural network structure with numerous layers of nodes (Multilayer Perceptron) sends input observations back and forth until it reaches a termination condition. ANNs were developed as pattern recognition and data analysis tools replicating the brain's neural storage and analytical functions. Nonparametric ANN approaches, unlike statistical classification methods, do not require prior knowledge of the input data distribution model. ANNs have the advantages of parallel processing, estimating the nonlinear connection between input data and desired outputs, and generalizing quickly. According to multiple earlier research on the classification of multispectral pictures, ANNs outperform standard classification approaches like maximum likelihood classifiers in terms of classification accuracy [41].

3.3.9 Parameter based

Metric approaches such as Support Vector Machines (SVM), Random Forests (RF), and Artificial Neural Networks have all been examined with per-pixel picture order projects (ANN). Well-known grouping strategies have been investigated employing spatial unearthly organization procedures, such as object-based picture inquiry, with significant increases in characterization exactness (OBIA). OBIA-based research, on the other hand, has been limited to the VHR/HR picture files, which are only accessible to specialists [42].

3.3.10 Clustering-based classification

Without human intervention, clustering is an unsupervised machine-learning job that splits data into clusters or groups of related things. It accomplishes this without informing how the groups should appear ahead of time. The technique of putting related elements together is known as "clustering." This unsupervised machine learning approach looks for commonalities in data points and groups them [43].

3.4 The classifiers of ML for LULC

3.4.1 The supervised classifiers

Algorithms that 'learn' patterns in data to predict a discrete class are known as supervised classification approaches. Machine learning techniques are a collection of flexible statistical prediction approaches. The supervised classification use of training data considered representative of each parameter type or unit to be classified is referred to following supervised classifiers:

3.4.1.1 Random Forest –RF

One of the better methods for classification is the RF algorithm. RF is capable of accurately classifying large amounts of data. It is a learning system in which many decision trees are built during training, and the individual trees anticipate the modal outputs. RF is a compilation of Classification and Regression Trees created via discretionary resampling on the readiness set using datasets of equal size to make up a set known as bootstraps. Many bootstraps are used when a tree is built as the test set to avoid joining a specific record from the first dataset. The theory botch as employing a test set of equivalent size as the arrangement set is a measure of the botch speed of the plan of all the test sets. The standard eliminates the need for a different test set. Each tree's hidden branches vote for one of two classes, and the forest forecasts which class will receive the most votes [44], as described in **Figure 6**.

3.4.1.2 Tree Decision Classification: TDC

A decision tree is an informative model gathered into a decision tree and has center points and constructed edges. The center links two inner issues: leaf center and leaf

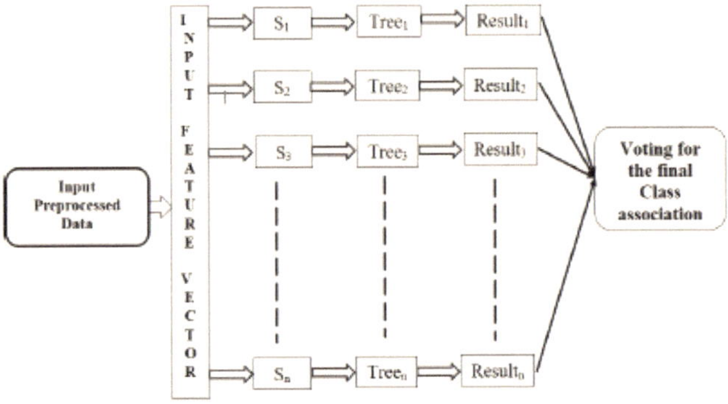

Figure 6.
RF system architecture [13].

center points. An inside center addresses a portion of the property, whereas a leaf center addresses a class mark. **Figure 7** depicts the planned path from the internal root center to the leaf center, which addresses the request, the fundamental standards, and the gathering measure utilizing decision tree regions as you'd expect from a rule-based classifier. The decision tree is straightforward to comprehend and unravel. It may be combined with various decision techniques to form an outfit learning classifier, such as a self-assertive woods classifier. Observe [39] that the difference between TDC & RF is apparent in **Table 4**.

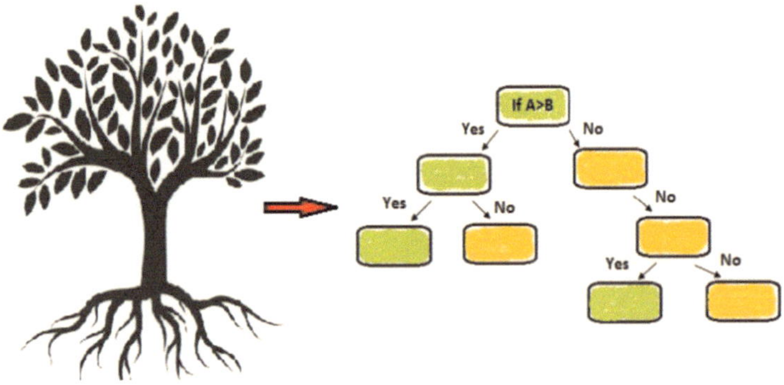

Figure 7.
TDC architecture.

Decision trees	Random forest
1. When decision trees grow unchecked, they frequently suffer from overfitting.	1. Overfitting is avoided because random forests are generated from subsets of data, and the final output is based on average or majority ratings.
2. A single decision tree is more efficient in terms of computing.	2. In comparison, it is slower.
3. A decision tree will create specific rules for predictions when given a data set with characteristics.	3. Random forest randomly selects data, creates a decision tree, and averages the outputs. It is not based on any formulas.

Table 4.
The difference between TDC & RF.

3.4.1.3 Maximum Likelihood Classifier: MLC

It is one of the most often used remote sensing classification algorithms, in which a pixel with the highest probability is categorized into the appropriate class [40]. It is used in distant detecting order applications. The maximum likelihood classifier MLC computation necessitates proper agent preparation of test information for each category and a detailed assessment of the mean vector and a covariance grid. MLC is a parametric classifier that addresses the inconstancy of courses by applying the covariance grid based on the likelihood that a pixel belongs to a given class. MLC may produce better results than other known parametric orders [37].

3.4.1.4 Spectral Angle Mapper Classification: SAM

It's a method of comparing photo spectra to a specified range or an automated end member, usually done with a spectrometer in a lab or the field. According to SAM, the data has been reduced to apparent reflectance, according to SAM [41]. **Figure 8** shows a technique called Spectral Angle Mapper that maps spectral angles. In an n-D space, where n is the number of bands, this method calculates the spectral angle between both (the unknown and known) spectra as vectors. The size and orientation of each vector are unique. The length of the vector represents the Pixel's brightness, whereas the vector's direction represents the Pixel's spectral characteristic [42].

3.4.1.5 Support Vector Machine: SVM

SVM is a quantifiable learning Theory-based conditional artistry grouping algorithm. This method is intended to be independent of the dimensionality of the component space [42]. The basic idea behind this arrangement is to use limited pixels to make a choice plane that isolates the classes by enhancing the edge between them. A choice plane separates a group of articles with varying levels of class involvement. The chosen planes may not necessarily be straight lines, as many characterization projects make this impracticable. Tasks that attract separate lines to distinguish various things are known as characterization tasks. **Figure 9** shows the situation [42].

3.4.1.6 K-Nearest Neighbor: KNN

KNN was a nonparametric approach utilized in statistical applications. The main idea behind KNN is to discover a collection of k samples in the calibration dataset that are the most comparable to unknown models (based on distance functions, for example) [43]. The response variables (i.e., the class characteristics of the k nearest neighbor) from these k samples are averaged to establish the label (class) of unknown data.

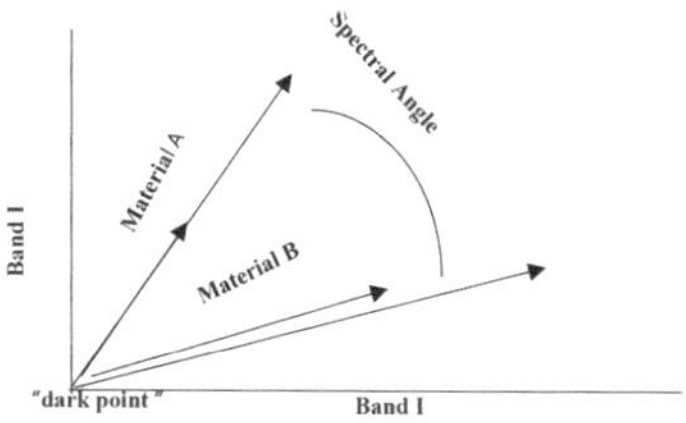

Figure 8.
Spectral angle mapper (SAM).

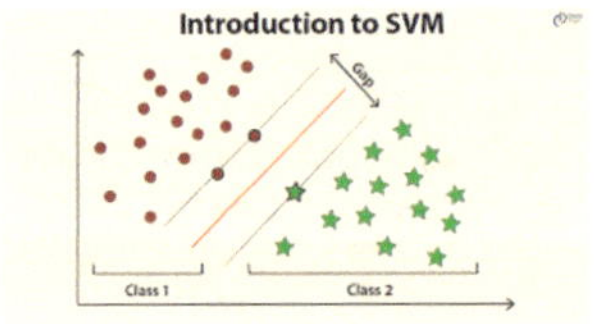

Figure 9.
Support Vector Machine (SVM).

As a result, the k plays a critical role in the KNN's performance for this classifier and is the most vital tuning Parameter for the KNN. A bootstrap technique was used to calculate the parameter k [38].

3.4.1.7 Minimum Distance Classification: MDC

As a regulated arrangement, the spacing between pixels is in the highlight space. It is commonly thought that highlight focuses of the same class are grouped in component space. The mean vector governed by this element focuses as the class's focal point, and the covariance network represents the scattering of encompassing focuses [39]. Every type has its own set of estimates for focus. The basic premise of the similitude measure is that if the highlight contrasts of the two modes are below a given edge, the models should be comparable [40].

3.4.1.8 Artificial Neural Networks: ANN

ANN Classification is learning to divide data into multiple groups by identifying common characteristics across samples from different classes. ANN of Supervised Learning Classification. Known class labels aid in determining whether or not the system is operating correctly [37]. Information, hidden, and yield layer make up its strategy [41]. The neuron receives the contribution from the left, and each piece of information is multiplied by a weight factor. Learning occurs when the loads in the hub are changed to reduce the gap between the yield hub actuation and the yield [42].

3.4.1.9 Mahalanobis

The Mahalanobis distance is a distance classifier that is sensitive to direction. For each form of input data, it utilizes statistics. While Mahalanobis distance is comparable to maximum likelihood classification, it is quicker since it assumes all class covariance is the same. Because no precise distance cutoff value was applied during software processing, the approach could identify all pixels to the nearest training data [43]. The Mahalanobis distance is a useful multivariate metric for determining between two points. It's a helpful statistic with applications in multivariate anomaly detection, severely unbalanced dataset classification, and one-class classification. Mahalanobis remote learning has sparked considerable interest, see **Figure 10**.

3.4.1.10 Maximum entropy

Maximal Entropy is a group method to the entropy selection criterion that was first proposed. The ensemble classifier's predictions are used in this method. The

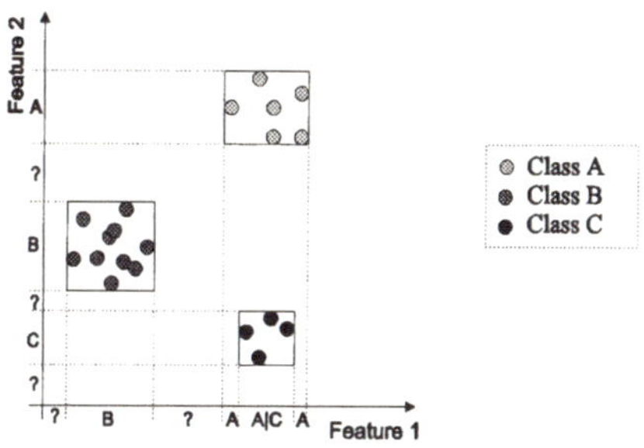

Figure 10.
Mahalanobis architecture.

collection's greatest Entropy determines the estimated Uncertainty measure for one instance. (Top) Bounded domain constraint x ∈ [0.7, 1.3] for the traditional equilibrium entropy S eq = ln p eq (x) eq which gives a flat profile and the trajectory entropy S FIT = ρ|∇ 2 |ρ which gives a distribution that scales as p * FIT ∼ cos 2 ((x − μ)π/2 L). (Bottom) Maximum entropy distribution under the constraint on the average x = μ and variance (x − μ) 2 = σ 2 which are equivalent for the static and trajectory information measure as p * FIT = p *, as cleared in **Figure 11** [38].

3.4.1.11 Parallelepiped

A fundamental decision method is used in parallelepiped classification. The decision boundaries in an image data space form an n-dimensional parallelepiped. A standard deviation threshold from the mean for each selected class determines the dimensions of a parallelepiped classifier in **Figure 12** [39].

3.4.1.12 Normal bayes

A Bayesian classifier's learning module creates a probabilistic model of the characteristics and uses it to predict the classification of a new example. The vector may be used to train a Bayesian classifier. The training data can compute the covariance matrices of the discriminant functions for the abnormal and normal classes [41]. Instead of calculating the maximum of the two discriminant functions, abnormal(x) and standard (x), the choice was made based on the ratio gabnorm(x)/standard (x).

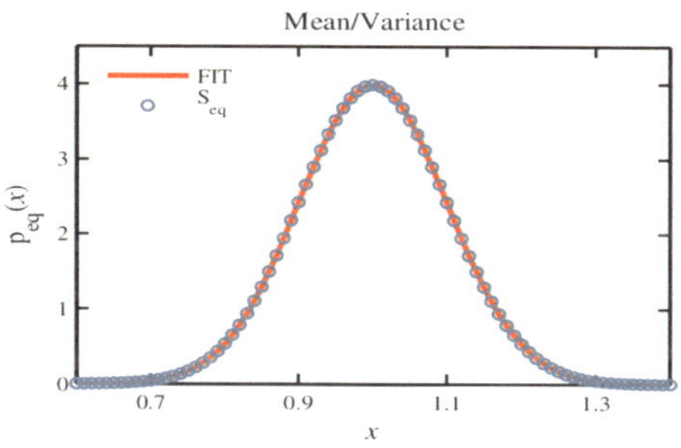

Figure 11.
Maximam entropy.

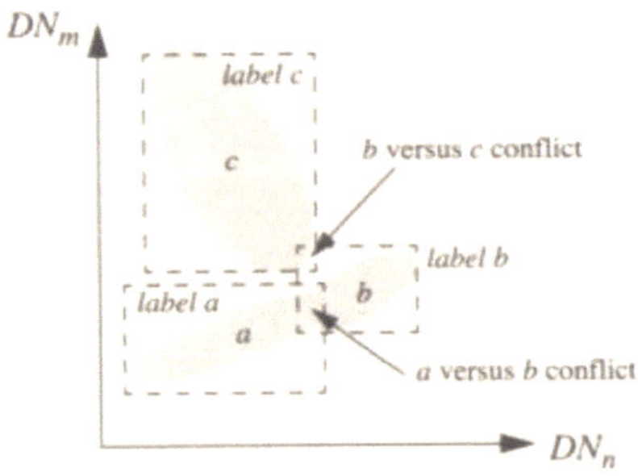

Figure 12.
Parallelepiped.

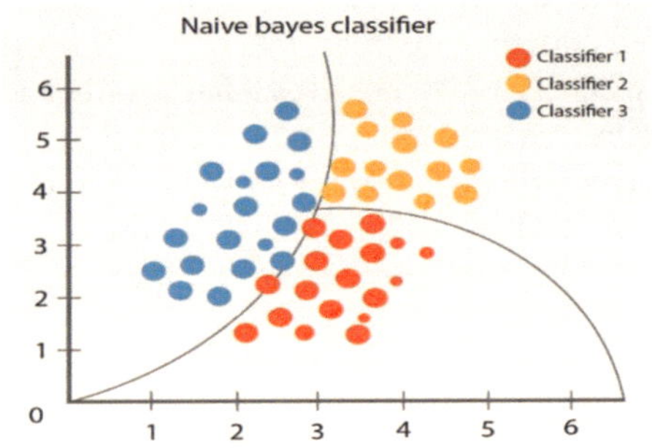

Figure 13.
Bayesian classifier.

The unknown pattern vector is categorized as odd if the ratio is more significant than T; it is expected, as shown in **Figure 13** [40].

3.4.2 Unsupervised classifiers

Based on geophysical response similarities, unsupervised classification algorithms can objectively classify anomalies into potentially relevant subsurface classifications. Unsupervised classification tries to classify pixels in a remote-sensing image into groups with similar spectral properties without human intervention. Several statistical techniques known as "clustering," which forms classes of pixels based on their shared spectral signatures, are used to create variety. The following are supervised classifiers [40]:

3.4.2.1 ISOData

It establishes equally distributed class means throughout the data space, then repeatedly clusters the remaining pixels using minimum distance algorithms. Every cycle, the means are recalculated, and the pixels are reclassified. Unsupervised classification with ISODATA calculates class means evenly distributed in the data space, then clusters the remaining pixels using minimum distance approaches. Every iteration recalculates the means and reclassifies pixels based on the new means. The ISODATA algorithm is an iterative method that clusters data components into different classes using Euclidean distance as the similarity measure [37].

3.4.2.2 K-means

It iterates until the best centroid is obtained by calculating the centroids. The data points assigned cluster in the method, resulting in a minimum total squared distance between data points and the centroid. The K-means clustering technique is utilized to locate groupings that have not been explicitly identified in the data [41].

4. Applications for land observation satellites

4.1 LULC change observation

Understanding LULC is essential for managing land resources and evaluating the potential technology [43]. Policymakers can use LULC change detection to understand environmental change dynamics better and assure long-term growth. As a result, LULC feature identification has become a hot topic in research, demanding the creation of a solid and reliable LULC classification method. Land use cover is required to make up the land's physically existent and apparent surface components [44]. LULC data are necessary for some planning and administration activities, and it is a critical component for illustrating and comprehending the earth as a system. It also plays an essential role in earth-atmosphere interactions [45].

LULC items in any location are essential as a natural and socio-economic component. LULC objects are deep data for various developmental activities on the earth's surface and their application to human needs in time and space. The land is crucial for humans to carry out any development activities on the planet's surface, such as agriculture, settlements, and industry. LULC information in the form of maps and statistical data is beneficial for studying land cover patterns, such as agriculture, forestry, economic production, settlements, and environmental studies for spatial planning, management, and land use and exploitation [46]. The LULC classification, without a doubt, plays a critical role in the regional socio-economic development of countries and the management of natural resources, **Figure 14**.

The LUCC study's applicability may be used to develop sustainable development in vegetation changes, quantity and quality of water resources, land resources, and coastal management. LULC maps play a crucial and pivotal role in arranging executives and monitoring initiatives. The importance of the discovery of the change of land use land cover LULC based on remote sensing data is the source of information to make appropriate decisions for the benefit of the countries. Disclosure of land change is a factor in conserving land and considering management and development [38]. LULC statistics are required for planning, business, and regulatory needs. The information is also essential for ecological security and spatial arranging because of its spatial nuances. Land use arrangement is indispensable because it gives information that can be utilized to demonstrate, particularly the one managing climate. For example, models manage environmental change and strategies improvements [39].

Land use research and analysis have become prerequisites for proposing a region's formative activities. In many developing countries, land assets form the foundation for financial development at the national, regional, and local levels. Land usage and land cover data are essential for organizers, decision-makers, and those concerned with land asset management [39]. It enables researchers to look at landscape patterns and features crucial to understanding land size, location, and condition, as well as the ecosystem's size, structure, and state. Land classification is essential because the usage

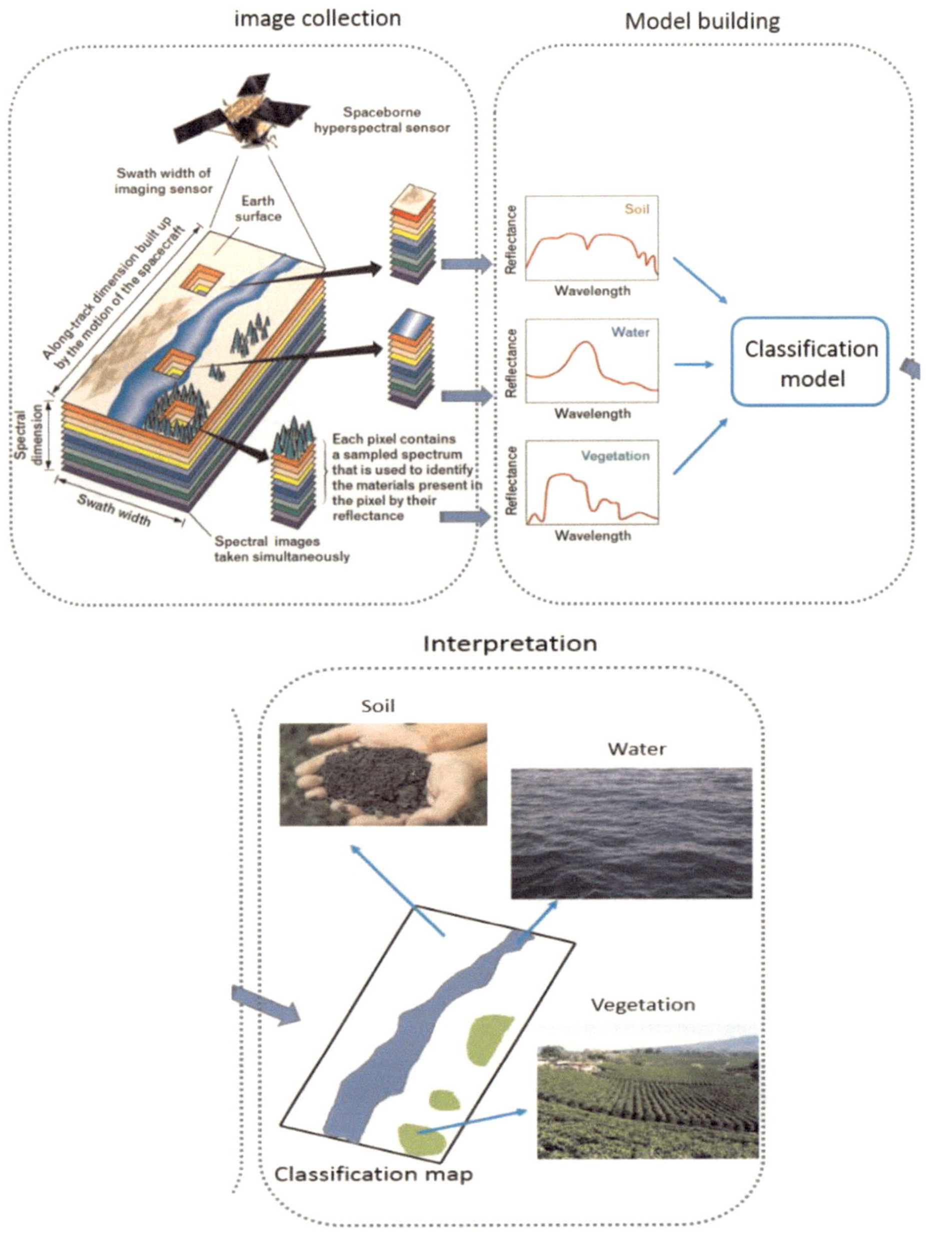

Figure 14.
Needing to Study LULC.

of a particular piece of land might be connected to considerable price disparities, necessitating a well-defined land categorization. For example, the price (development) of land beneath dwellings may differ significantly from the price (impact) of agricultural land [40]. It provides essential information about human use of the terrain.

Land classification is essential because the usage of a particular piece of land might be connected to considerable price disparities, necessitating a well-defined land categorization. For example, the price (development) of land beneath dwellings may differ significantly from the price (impact) of agricultural land [37], where land use research and analysis have become prerequisites for proposing a region's formative activities. In many developing countries, land assets form the foundation for financial development at the national, regional, and local levels. Land usage and land cover data are essential for organizers, decision-makers, and those concerned with land asset management [41].

4.1.1 Materials and procedures

4.1.1.1 Sana'a study area

Sana'a is one of the largest cities in Yemen and is in the governorate of the same name as well, and this city is the case study for this article [31]. The city of Sana'a is located at 15°N 44°C or 15.369445 latitudes 44.1191006 with 15°22′ 10.0020'N and 44° 11′ 27.6216″E in GPS coordinates [42].

The city of Sana'a Total area is 126 km2 (49 sq. mi), and the population was 2,545,000 issued in 2017. The city has an environment of around 2200 meters above ocean level, see **Figure 1**. The north-central part of Yemen it's in a high valley that runs from south to north [12]. With an entire space of 126 Km2 (49 sq. mi), it has a populace of around 3,937,500 (2012). Sana'a's precipitation is limited to 200 mm/year, while the fading is several times higher.

The average daily sunlight-based irradiance ranges from 800 to 1400 μmol/m^2, with the month-to-month average air temperature between 22 and 30°C at low humidity levels (35–55%). Its climatic conditions (temperature, sun-based radiation) are ideal for wastewater treatment based on phototrophic [43], as described in **Figure 15**.

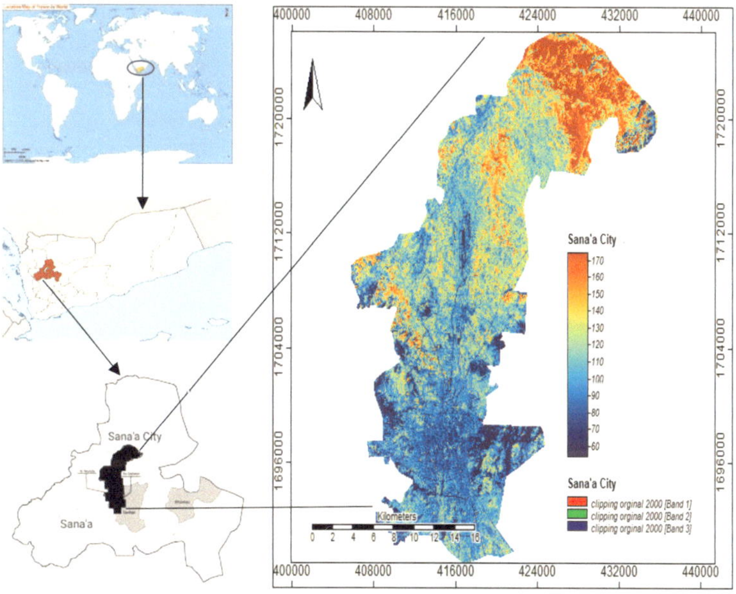

Figure 15.
The location of Yemen Country in the world, Sana'a Governorate, and Sana'a City.

4.1.1.2 Satellite data

This article used the Landsat8 Satellite Sensor (30 m) for LULC mapping & geometrically open-source Landsat8 MSS/TM. The image was obtained from the United States Geological Survey (USGS) of the Sana'a region, a scientific body of the US government. The base map was created from survey photos of the SOI toposheet at a scale of 1:50000 [44]. In this study, the data collected in 1980, 1990,2000,2010,2020 the database details created in **Table 5**. You can see Sana'a Region on google Maps in **Figure 16**. The data set of Landsat8 Satellite Sensor (30 m) capture & selection area study with Composite band 432 in **Figure 17**.

4.1.1.3 Methodology

The following diagram illustrates the essential steps of this research study in **Figure 18**.

No	period	Satellite	Sensor	Resolution
1	1980	Landsat 5	(TM)	30 m
2	1990	Landsat 5	(TM)	30 m
3	2000	Landsat 7	(ETM+)	30 m
4	2010	Landsat 5	(TM)	30 m
5	2020	Landsat 8	(OLI), (TIRS)	30 m

Table 5.
Database created of images LULCC of Sana'a city.

Figure 16.
Sana'a Region in google map.

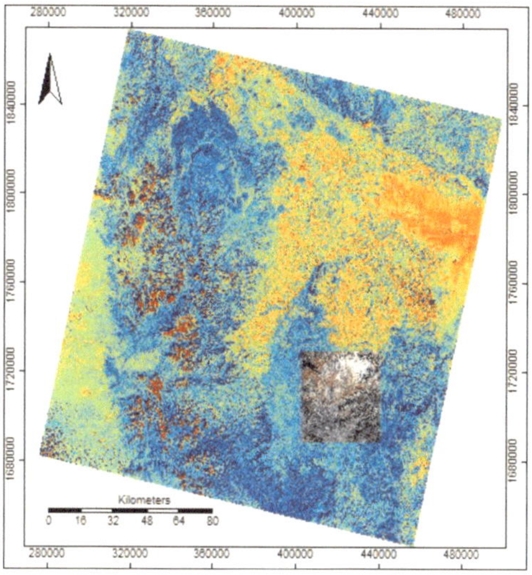

Figure 17.
Data set of Landsat8 with Composite band 432.

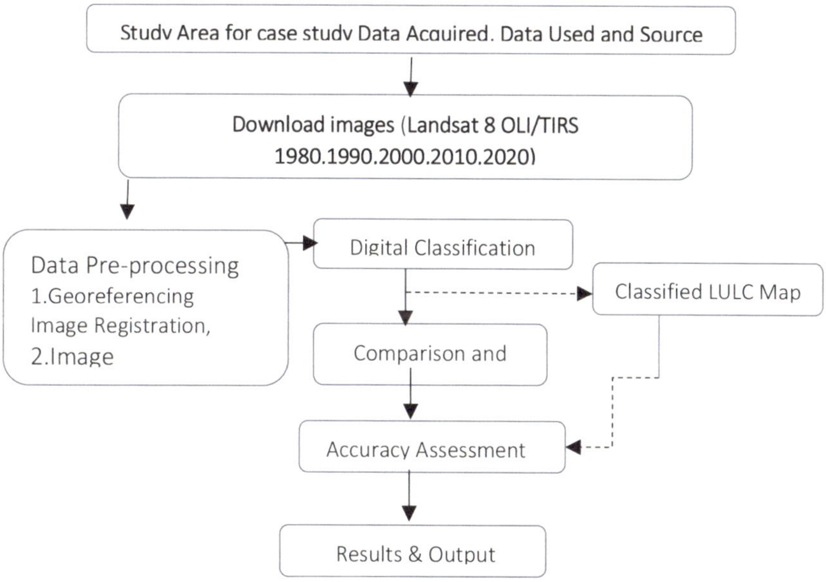

Figure 18.
Workflow diagram for proposed methodology.

4.1.1.4 Create database

For constructing a database of observation land changes of Sana'a' in Yemen, the data used in the LULC classification here are 1980,1990,2000,2010,2020 for extracting the differences of decadal period land changes of the region, and the composition of the database is shown in **Table 5**. Note Operational Land Imager (OLI) and Thermal Infrared Sensor (TIRS), Thematic Mapper (TM) Enhanced Thematic Mapper Plus (ETM+), and Multispectral Scanner (MSS).

4.1.1.5 Pre-processing

It is the primary stage and essential task in the process of LULCC, the coordinate reference system for defining and cutting the map into specific areas. The pre-processing procedure identifies the data after it is downloaded from satellites under remote sensing technology. The information subject to pre-processing is divided into the images shown in WGS84 or WGS84 / UTM.

They are pre-processed to contain valid data with a geometrically calibrated reflection often present in the upper atmosphere. The data is not overtly distributed, but its implementation is terminated by organizations responsible for managing satellites. **Figure 7** will display the pre-processing corrections for Landsat 8 satellite images in which Band 543 in decadal time 1980,1990,2000,2010,2020.

The images cleared differences in these images map before classification. According to the colors of the Landsat satellite, the region's red color is vegetation, and the white color is bare land, light gray is land area & network road, and dark gray is built-up area. Initially, the comparison is clear how to land in Sana'a city is changed. The data set of Landsat8 Satellite Sensor (30 m) & selection area study with Composite band 432 cleared in **Figure 19**.

4.1.1.6 Classification for Sana'a city land from 1980 to 2020

As input layers for model processing, there are six samples for six parameters for creating model classes: High Land, Mountains, Land Area, Built-up, Vegetation, and Bare Land. Note to parameters in software SAGA with these models classification in down are seven, but in processing and results in the Parameter are six since merge area vegetation with agriculture land. Create the samples depending on RGB color composites of sentinel-2A images, for example, the class Vegetation (red pixels in color composite RGB = 432), detailed changes in the region. The following details illustrate the critical description of class input in **Table 6**.

4.1.1.7 Land changes

Figures 20 and **21** images indicate geomorphological changes in Sana'a in the recent period. After 2010, that change has a role in analyzing this study. This study showed the differences in geomorphology during the mentioned period through the land change classification, which suggests that land use in this region is inappropriate. A database of LULC of Sana'a was created in this work. RF classifier used with Landsat images satellites.

Such research is necessary for developing nations because it will aid in managing natural resources, where LUCC plays a critical role in regional economic development and natural resource management. Destroyed the country's infrastructure, preventing Sana'a's vital economic, social, environmental, health, and agricultural development.

This study shows the detailed Analysis classification for Sana'a city land from 1980 to 2020. LULCC was done in 1980, 1990,2000,2010,2020. It can find LULC classified for Sana'a city, and the categories can be apparent in the differences in land change in Sana'a city as shown in **Figure 22**.

The summary report compares built-up Areas and Land areas through 1980, 1990,2000,2010,2020. The findings revealed that the political problem began after 2010, as the built-up area decreased on a map in 2010 while the land area increased. I was implying that the poor state of Sana'a city was caused by the war, with increasing

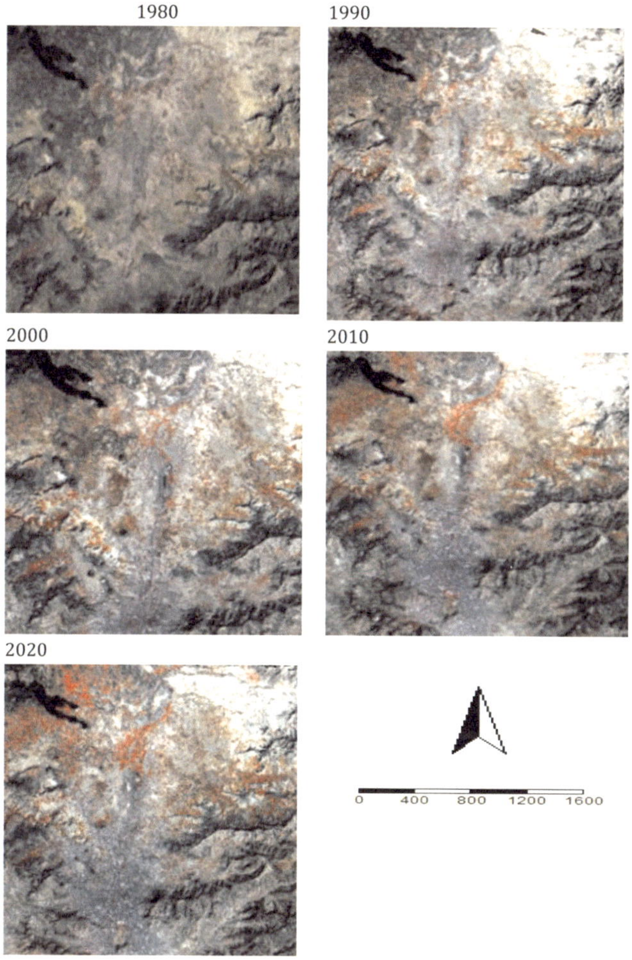

Figure 19.
Data set of Landsat8, selection area study with composite band 432.

LCLU Class	Description
High Land	High Land Remote may be settlements and clans with a long history and profound loyalties.
Mountains	A mountain is a raised section of the earth's crust with steep sides and exposed bedrock.
Land Area	The area in square kilometers of the land-based portions of conventional geographic regions is called the land area, which is the population of people. Not contains buildings, maybe streets, parks, roads or buildings crashed down, like this.
Builtup	Built-up areas may be Large buildings, small buildings, settlements, transportation, land, or places containing people like banks, schools, hospitals, etc.
Vegetation	Space containing crops, fields, sparse grassland, a Temperate steppe, and a Temperate meadow.
Bare Land	Bare soil, bare rocks, and land do not contain people like the desert.

Table 6.
Description of LULC classes in the study area.

Figure 20.
Buildings & infrastructure of Sana'a city before the conflict [3].

Figure 21.
Post-conflict images show the change of the built-up land to the destroyed land of Sana'a's city [35].

built-up area in town resulting in a decrease in land area and the decreasing built-up area in the city's growing land area. **Table 7** and **Figure 23** show the opposite situation.

The summary report is apparent in **Figure 23** to compare built-up Areas and Land areas through 1980, 1990,2000,2010,2020. The findings revealed that the political problem began after 2010, as the built-up area decreased on a map in 2010 while the land area increased. Increasing built-up area in town results in a decrease in land area and the decreasing built-up area in the city growing land area.

4.1.1.8 Results

According to the study's findings, **Table 8** shows the area and percentages of LULC over the decadal period of Sana'a City from 1980 to 2020. All the region's size factors have been displayed: high land, mountains, land area, built-up area, and vegetation. According to the findings of this study, the built-up area in 1980 was 12.17 percent, and it rose by 34.24 percent in 1990. That is typical, and expansion will continue because of human activity in front of increasing structures and urban development.

The built-up area was 40.15 percent in 2000, then decreased to 30.94 percent in 2010, which is not typical. The built-up area was 44.74 percent in 2020. Perhaps this is back to political events after 2010 that led the development movement backward in all sectors, including the economy. The remaining analysis parameters had an impact on increasing and decreasing.

The destruction of missiles and the expansion of barren terrain are the main reasons for the shrinking built-up area. The results of land change are mentioned in detail in **Figure 24**. The area under significant land-use or land-cover classes was calculated for 1980,1990,2000,2010 and 2020. The region's area in 1980 was 1,867,950,000 km^2, and in 2020 was 1,497,207,600 Km2. The difference between

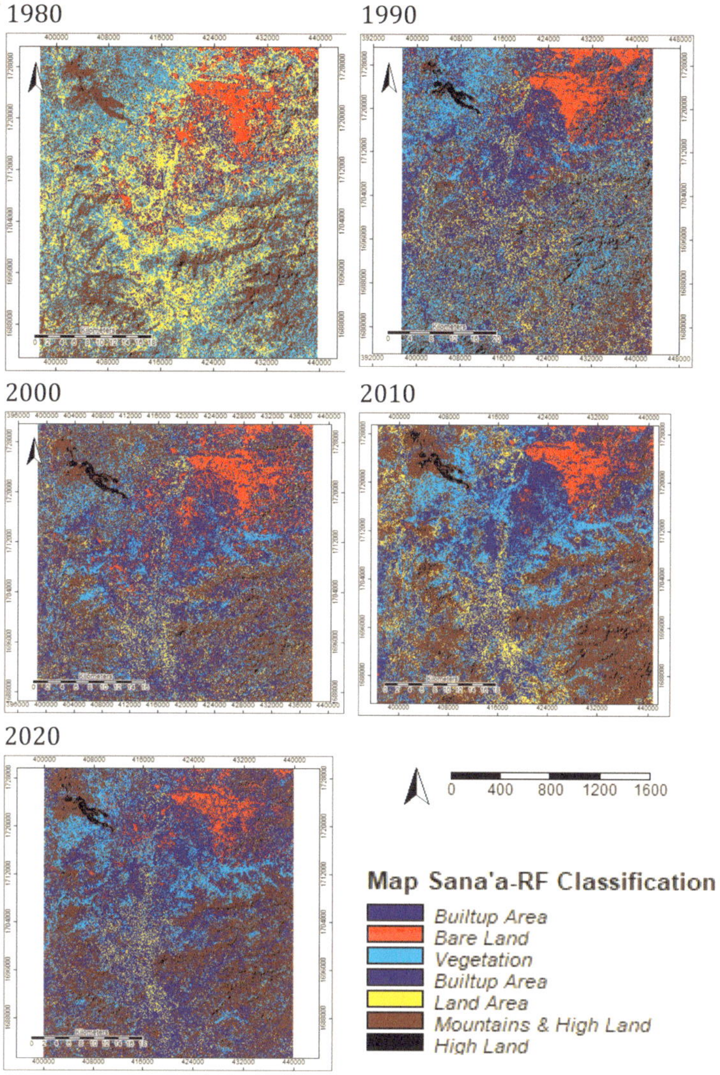

Figure 22.
Classified map for Sana'a over the five decades (1980–2020).

Year	Builtup Area km^2	%	Land Area km^2	%
1980	21,276	9.99%	50,965	48.14%
1990	52,590	24.69%	20,428	19.29%
2000	51,041	23.96%	95,706	9.04%
2010	40,097	18.83%	17,404	16.44%
2020	47,980	22.53%	75,069	7.09%
Total	21,298,644	100.00%	1,058,762	100.00%

Table 7.
Results are calculated for class category land Area & Built-up from 1980 to 2020.

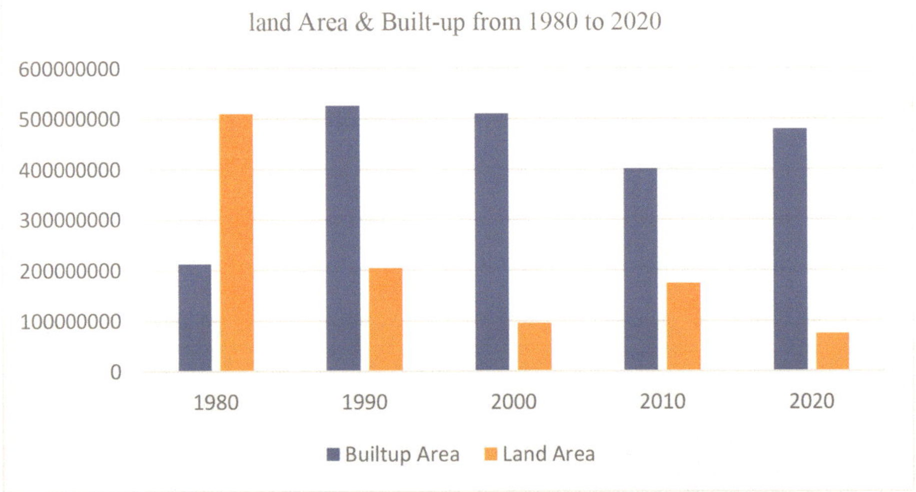

Figure 23.
The chart of results is calculated for class category land Area & Built-up from 1980 to 2020.

them is 370,742,400 km^2, which means the percentage difference is 19.85% of all geographic space of the city. During this period, there has been a persistent reduction in land cover as woodlands expand in cropland and developed regions. Somewhere in the range of 1980 and 2020, of the six significant LULC classes, an extensive increase and decrease have been recorded see **Table 8**.

The study concluded that human factors and processes have greatly affected the shapes of the earth's surface in Sana'a by comparing maps for the years 1980 and the year 2020. Human activities have affected the disappearance of many forms of the earth's surface that contain gains from the Yemeni civilizational heritage, such as castles, forts, and caves, due to human activities and the work of crushers in the mountains. It was reached to create a database for a geomorphological map of the study area. The study recommends valuing biological and human geographical studies to identify the processes and factors affecting the formation of the earth's surface forms. They benefit from planning and conducting comprehensive development projects and employing them to develop the mountainous heights in Sana'a through building dams and parks and establishing a shelter. The importance of Benefiting from the study of spatial analysis and choosing the optimal site through geographic information systems to make service projects, such as planning to establish a water barrier.

4.1.1.9 Discussion

The study's findings showed the area and percentages of LULC over the decadal period of Sana'a City from 1980 to 2020. All the region's size factors have been displayed: high land, mountains, land area, built-up area, and vegetation. According to the findings of this study, the built-up area in 1980 was 12.17%, and it rose by 34.24% in 1990. That is typical, and expansion will continue because of human activity in front of increasing structures and urban development. The built-up area was 40.15% in 2000, then decreased to 30.94% in 2010, which is not typical. The built-up area was 44.74% in 2020. Perhaps this is back to political events after 2010

No	NAME	1980		1990		2000		2010		2020	
		AREA m^2	%	AREA m^2	%	AREA m^2	%	AREA m^2	%	AREA m^2	%
1	High Land	171,612	0.84%	932,976	3.64%	316,881	1.83%	492,039	2.41%	485,370	2.53%
2	Mountains	5,788,260	28.34%	6,728,850	26.23%	5,127,039	29.61%	7,512,750	36.75%	5,990,517	31.24%
3	Land Area	5,096,592	24.95%	2,042,865	7.96%	957,069	5.53%	1,740,402	8.51%	750,699	3.92%
4	Builtup Area	24,867,360	12.17%	8,784,819	34.24%	6,952,491	40.15%	6,325,785	30.94%	8,578,098	44.74%
5	Vegetation	5,334,408	26.11%	5,383,881	20.99%	2,224,908	12.85%	3,333,969	16.31%	2,307,519	12.04%
6	Bare Land	1,549,656	7.59%	1,779,786	6.94%	1,735,965	10.03%	1,037,925	5.08%	1,060,686	5.53%
7	Total of area	20,427,264	100%	25,653,177	100%	1,731,435	100%	20,442,870	100%	19,172,889	100%

Table 8.
Area and percentages LULC for decades period of Sana'a City from 1980 to 2020.

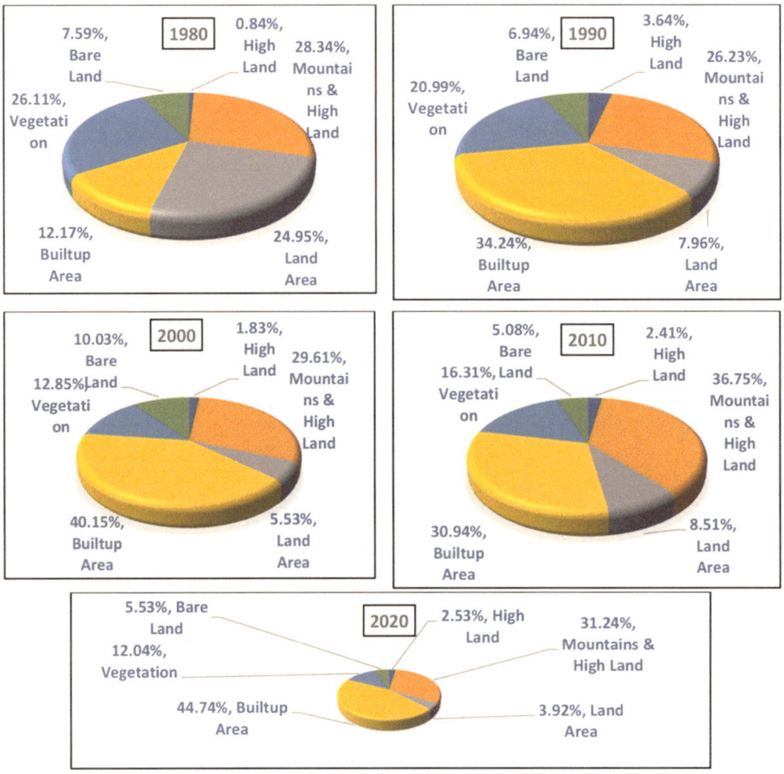

Figure 24.
Percentages LULC for decades period of Sana'a City from 1980 to 2020.

that led the development movement backward in all sectors, including the economy. The remaining analysis parameters had an impact on increasing and decreasing.

Regarding Sana'a, the region has progressed in urban density, built-up area, and bare ground before 2010 and the opposite after 2010. The area under significant land-use or land-cover classes was calculated for 1980, 1990, 2000, 2010, and 2020. The region's area in 1980 was 1,867,950,000 $m^2$and in 2020 was 1,497,207,600 m^2; the difference between them is 370,742,400 m^2, which means the percentage difference is 19.85% of all geographic space of the city. During this period, there has been a persistent reduction in land cover as woodlands with attending expansion in cropland and developed region. Somewhere in the range of 1980 and 2020, of the six significant LULC classes, an extensive increase and decrease have been recorded.

This study tried summary of the factors and reasons potentially of the land changes in the Sana'a region is as follows:

1. Events of the war, the asset verification revealed findings that the damage on the ground matches evidence shown in satellite and field photographs. Furthermore, satellite imaging can be used to quickly verify assets when analyzing large-scale damage [24].

2. Human factors for extraction of building materials. Building houses and residential buildings required different building materials, including gravel,

which led to the great demand for the development of many mines on those sites and the establishment of crushers that affected the shapes of the Earth's surface. They worked to drain critical natural resources, which shows human activities and operations in the production areas in the highlands.

3. Erosion and climate factors and Occasional flash floods and potential disasters. Several natural characteristics in several aspects characterize the study area. Climate, one of the biological factors, played a role in the formation and change of these manifestations. Human processes increase as the population grows, reflected in the number of people. Construction, quarrying, and other lands in agriculture and industry are examples of human operations and activities from 1980 to 2020, with an average consumption and conversion of building materials of about five tons.

4.2 Satellite altimetry monitoring changes in mean sea level

The marine gravity field is primarily reliant on satellite altimetry. The accuracy and resolution of the marine gravity field model have been significantly improved due to the development of altimetry missions and advancements in altimeter data processing techniques. However, recovering high-accuracy and high-resolution gravity fields from satellite altimeter data continues to be a difficult task.

Since altimeter data processing techniques are crucial for obtaining precise measurements of sea surface height, these upgraded approaches are then discussed and reviewed with a focus on coastal altimetry. The difficulties in processing altimeter data are also emphasized. The characteristics of gravity recovery methods, including least squares collocation, the inverse Vening Meinesz formula, the inverse Stokes formula, and the inverse Vening Meinesz formula, are also reviewed in the third section. The most recent global marine gravity field models, altimeter data, and processing methods are also shown.

Shipboard gravity measurements also assess the effectiveness of the current global gravity field model. In the low-middle latitude regions, the root means square of the difference between the shipboard gravity from the National Centers for Environmental Information and the global marine gravity model is roughly 5.10 mGal, which is better than the outcome in high-latitude regions. The accuracy of models in the coastal areas still has to be improved, especially within 40 km of the coastline. The SDUST2021GRA model created by the Shandong University of Science and Technology team also showed a fascinating performance. The difficulties in recovering the marine gravity field from satellite altimetry are finally discussed [46].

The sensor aboard an altimetry-focused satellite delivers microwave pulses in the radar frequency range or laser pulses in the optical or infrared spectrum to the ground, which is reflected at the planet's surface and collects the return signals. The radar observation method is particularly suitable over oceans and open water on land due to the favorable, reflecting characteristics of water. The fundamental measurement is the signal's round-trip time from the satellite to the water's surface, which, when multiplied by the speed of light, equals the signal's round-trip distance. The range, or separation between the satellite and the instantaneous sea surface, is roughly equal to the product of the two-way travel time and the speed of light.

The satellite's height above a global ellipsoid is calculated from its orbit about a geocentric reference frame (e.g., the International Terrestrial Reference Frame – ITRF).

A laser altimeter, which works similarly to radar technology but employs light pulses, can also determine altitude. Digital elevation models are frequently created using laser altimetry and measuring the elevation change of ice sheets. Their mass balance in response to global warming, satellite radar, and laser altimetry has more recently been used to measure the water level of lakes, rivers, and floodplains on land. **Figure 25** indicates Satellite Altimetry Monitoring Changes In Mean Sea Level [48].

Sea surface height (SSH) readings from satellite altimeters are a regular source of information for tracking ocean processes. It is difficult to completely utilize the available altimeter observations to correctly examine minor mesoscale variations in SSH because, below a wavelength of about 70 km, along-track altimeter measurements frequently exhibit a severe decline in signal-to-noise ratio (SNR).

Although many different strategies have been put forth and used to separate noise from measurements and detect it, no transparent methodology has evolved for systematic use in operational products. The Copernicus Marine Environment Monitoring Service (CMEMS) offers detailed band-pass filtered data to reduce noise contamination of along-track SSH signals to best address this unresolved issue. Users looking to reveal small-scale altimeter signals are thus left to their own devices to devise more creative and appropriate noise-filtering solutions [49].

Here we show that an entirely data-driven strategy is effectively designed and deployed to produce reliable estimates of noise-free sea level anomaly (SLA) signals

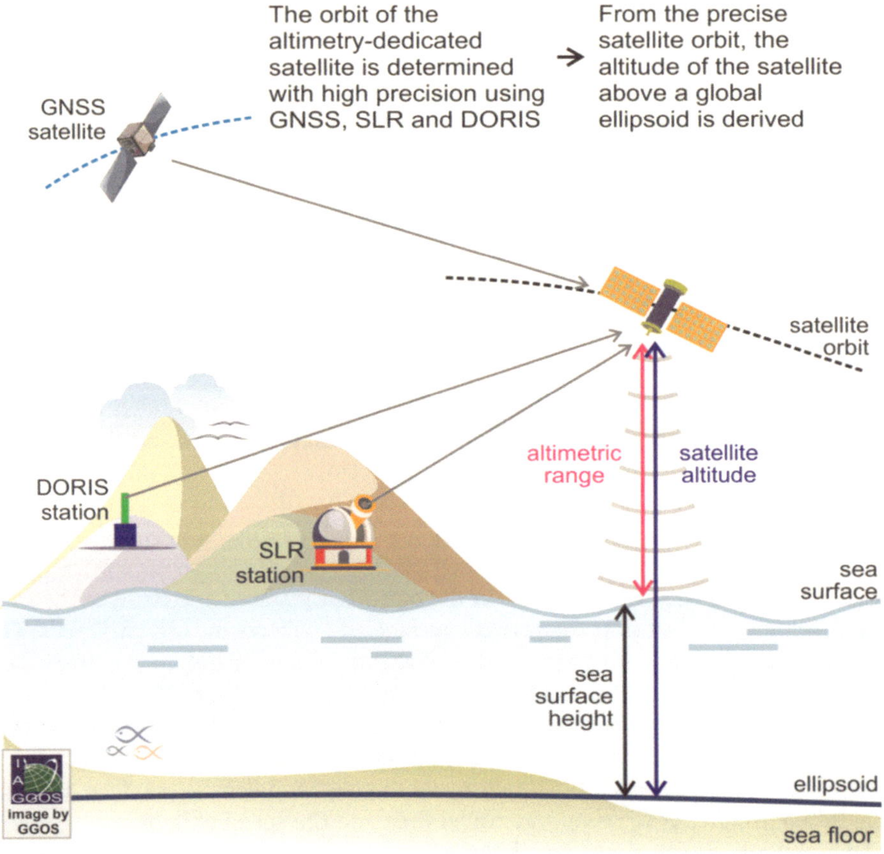

Figure 25.
Satellite altimetry monitoring changes in mean sea level [47].

(Quilfen, 2021). The approach combines a discrete wavelet transform (DWT)-inspired adaptive noise filtering technique with empirical mode decomposition (EMD), which is used to investigate non-stationary and nonlinear processes. It is discovered that this range of mesoscale wavelengths, between 30 and 120 km, better resolves the pattern of SLA variability.

The denoising method, which assumes that the SLA variability is partially the product of a stochastic process, results in a practical uncertainty variable associated with the denoised SLA estimations and considers errors related to the local SNR as well as process uncertainties. The measurements from the missions Jason-3, Sentinel-3, and SARAL/AltiKa are processed and analyzed. Their energy spectrum and seasonal distributions are defined in the small mesoscale domain over the period that is currently accessible. The SASSA (Satellite Altimeter Shortscale Signals Analysis) data set of denoised SLA measurements for three reference altimeter missions has already been shown to yield valuable opportunities to assess global small mesoscale kinetic energy distributions in anticipation of the upcoming SWOT (Surface Water and Ocean Topography) mission data (Quilfen and People, 2021) [47].

5. Challenges of AI techniques for LULC classification

Despite considerable recent progress in AI for LULC, global land-use intensity mapping has faced significant challenges in recent years. Artificial intelligence techniques have spread widely and provided many new solutions to various areas of the natural world and the difficulties of human society. On the other hand, the challenges of artificial intelligence techniques appeared for each field separately.

This section will identify the challenges of artificial intelligence techniques for mapping extensively, with challenges in monitoring how the land cover is classified. As platforms and sensors improve, new issues develop, such as high-dimensional datasets (high spatial resolution and hyperspectral features), sophisticated data structures (nonlinear and overlapping distributions), and the nonlinear optimization problem (high computational complexity) [29].

The complexity of multi-source data exacerbates the difficulty of developing robust and discriminative representations from training data with AI techniques [30]. It might be considered a diverse and significant data processing challenge. Large training samples are necessary for supervised AI systems, generally obtained through time-consuming and labor-intensive processes such as human interpretation of RS products and field surveys. With little training data, developing a robust model of AI-based approaches is a significant difficulty. Techniques for unsupervised AI must be developed.

There are a variety of AI models and frameworks that are both efficient and accurate. Researchers are continually proposing new AI-based change detection systems at the moment. However, it is a significant task to choose an efficient one and ensure its correctness for various applications. In practical applications, AI's dependability must be considered [31]. Some researchers have looked at these issues and suggested viable solutions, and it will summarize them separately as follows:

5.1 Issue AI's reliability

When using AI techniques for change detection, factors affecting the reliability of data preparation, model training, change feature extraction, and accuracy evaluation should be considered. The goal is to find the most plausible AI framework for enhancing change detection accuracy. We've discussed the issues and promises of AI-based change detection systems in this section and our forecasts for the future [32].

Although many AI-based change detection systems provide the model structure, their trainable parameters are opaque, making it difficult to comprehend why and how they work [33]. AI reliability aims to develop methods for improving the accuracy and interpretability of change detection systems. As a result, it is necessary to build change detection AI that is both resilient and interpretable [34]. **Table 9** describes only the approaches that can improve the accuracy of change detection findings from the following areas.

5.2 Issue AI without supervision

While domain knowledge can aid in constructing representations in classic machine learning methods, AI drives the development of more powerful unsupervised methods. Data can be used to teach unsupervised representation-learning algorithms

AI's reliability		
1	**2**	**3**
Reduce data uncertainty caused by geometric and spectral disparities by eliminating mistakes produced by data sources (such as preprocessing and radiometric correction) or merging different data to improve the original data's reliability, increasing change detection conclusions' dependability. A few studies have investigated the influence of registration and algorithm fusion.	Improve AI model interpretability by utilizing a sub-modular model structure, which can help comprehend the overall AI model's operation principle by understanding the role of each sub-module. R-region-proposals CNN's component, for example, can be regarded as a generator that predicts object regions.	Improve the durability of AI models by combining many approaches and outcomes. Ensemble learning is a good strategy for increasing the accuracy of the final output by combining the findings of multiple models.
4	5	6
Reduce noisy points and provide accurate boundaries by including post-processing methods such as the Markov random field, the conditional random field, and level set evolution into the AI model.	To improve the sharpness of change maps, use more suitable detection units. Based on the detection unit of change detection, it may be divided into scene level, patch or super-pixel level, pixel level, and sub-pixel level, from coarse to fine. In terms of dependability, the sub-pixel level is the best option because it avoids the problem of mismatched pixels in RS images. However, it swiftly escalates the level of computational complexity. As a result, the best solution is to employ separate detection units for different land cover types, which requires a well-designed AI model.	To improve the representation of change maps, detect changes in each instance. Change maps include binary, one-class mappings, from–to maps, and instance maps. Although research is still inadequate, the instance change map is more realistic. Because it can provide change information for each instance, it indicates real-world changes. It may also avoid the binary map's lack of semantic information and the classification system's restriction of the form–to map, increasing the dependability of the final result.

Table 9.
Directions for solutions issue AI's reliability.

to learn hierarchical properties. It's feasible to make data-driven decisions with it [35]. **Table 10** summarizes the aspects of unsupervised AI research.

5.3 Issue heterogeneous big information processing

Heterogeneity is a significant property of vast and heterogeneous data, and it complicates the formulation and analysis of change detection findings. SAR, GIS data,

Unsupervised AI		
1	2	3
Many researchers have not trained efficient AI models due to a shortage of labeled examples in recent years. I have put forth a lot of work to solve these issues and have regularly delivered excellent outcomes. GAN, transfer learning, and other unsupervised AI approaches are developing continually.	Change detection is frequently seen as a low-likelihood issue due to the ambiguity of the change location and direction (i.e., the unaltered change map is significantly more significant than the change). Unsupervised AI algorithms struggle to address this difficulty due to a lack of experience. Although more study is needed to improve performance, weakly- and semi-supervised AI systems are viable alternatives to supervised AI. Nonetheless, a pure unsupervised AI approach to change detection should be the ultimate goal.	One of the motivations for looking into unsupervised AI systems is the lack of training samples or prior knowledge. An excAn excellent alternative strategy is to use crowd-sourced data as a priori knowledge. The Web 2.0 age has arrived on the Internet (emphasizing user-generated content, simplicity of use, participatory culture, and end-user interoperability). For example, OpenStreetMap, a free wiki world map, may provide massive annotation data labeled by volunteers for AI model training. Although the label precision of specific crowd-sourced data is low, the AI model can be trained in a weakly supervised manner to detect changes.

Table 10.
Directions for solutions issue unsupervised AI.

Heterogeneous big data processing		
1	2	3
Several AI-based change detection systems based on heterogeneous data have proven effective. However, the number of studies is small. Furthermore, spotting changes across multiple data sources is more important to them than catching data fusion over time. Data fusion theories (i.e., mutual compensation of several types of data) and multi-source data (e.g., optical RS images and DEM) combined with AI techniques can help improve change detection accuracy.	Because current change detection methods rely mainly on detecting 2D data, using 3D data to detect changes in buildings and other structures is also a future research direction. Three-dimensional reconstruction based on oblique images or laser point cloud data and dimensional information integration based on aerial imaging and ground-level street view imagery (i.e., air-ground integration) is two. For recognizing 3D changes, there are currently no sound AI systems.	The application of the AI model is limited since analyzing RS's extensive data requires many CPU resources. Large-format data processing, for example, is frequently done in blocks, which might lead to edge concerns. A large amount of data needs many trainable parameters in the AI model, resulting in a lengthy training process that uses many computing resources. As a result, the volume of data and the number of trainable parameters must be balanced. They make developing AI-based change detection tools more challenging.

Table 11.
Directions for solutions issue heterogeneous big data processing.

high-resolution satellite pictures, and other time and space-measured data are just a few examples of the kind of data that RS technology may provide for change detection. These data of many sorts and formats are difficult to use due to missing values, considerable data redundancy, and unreliability. Furthermore, the generalization ability of current AI systems in RS data processing, particularly heterogeneous significant data processing, must be improved [36]. As a result, we believe the following issues warrant further examination, as in **Table 11.**

6. Conclusions

Classifiers that create exact LULC maps are in high demand, and dependable Information is required from remotely sensed pictures, even on high-dimensional, complex data. Machine Learning Classifiers have a significant role in giving good classification results. Several aspects influence the accuracy of classified maps, including training sample size, training sample quality, thematic correctness, classifier choice, study region size, etc. Understanding these criteria will aid in achieving the highest classification accuracy feasible for a given need. Big Data challenges arise when classification tasks involving multiple satellite photos and features become computationally intensive.

In recent years, artificial intelligence techniques have spread widely and provided many new solutions to various areas of the natural world and the difficulties of human society. On the other hand, the challenges of artificial intelligence techniques appeared for each field separately. This chapter identified the challenges of artificial intelligence techniques for mapping extensively with challenges in monitoring how the land cover is classified since advances in technologies catalyzed by machine learning and artificial intelligence.

One challenge is the infrastructure, especially the infrastructure of the ancient cities where roadworks were built at different times and with other materials. to face this challenge, it may help with multispectral data technology that can identify more objects and generate more categories. High resolution is the super-spectral for remote sensing data available for urban areas recently. And the accuracy challenge is among the significant challenges in this field. Where noticed was that many researchers used satellite imagery with an accuracy of 30 cm. Also, among the challenges, the challenge in image classification was the weak role of the analyst in the category and the possible classification errors.

The main challenge in LULC changes using remote sensing data to provide accurate and timely geospatial information is clarified as follows. Urban growth has long been considered a sign of regional economic vigor. Still, its benefits increasingly negatively impact the ecosystem and environment, including road traffic, air quality, loss of farming area, social fragmentation, and infrastructure cost. Natural resource management, planning, and monitoring programs depend on accurate information about the land cover in a region. The production of a thematic map from this classification using an image classification is one of the most common applications of remote sensing.

Acknowledgements

Thankful acknowledgments to team IntechOpen for supporting my achievement in this chapter.

DOI: http://dx.doi.org/10.5772/intechopen.110039

Author details

Eman A. Alshari[1,2]* and Bharti W. Gawali[2]

1 Computer Science and Information Technology Department, Thamar University, Dhamar, Yemen

2 Dr. Babasaheb Ambedkar Marathwada University, Aurangabad, India

*Address all correspondence to: em.alshari3@gmail.com

References

[1] Bioucas-Dias JM, Plaza A, Camps-Valls G, Saunders P, Nasrabadi N, Chanussot J. Hyperspectral remote sensing data analysis and future challenges. IEEE Geoscience and Remote Sensing Magazine. 2013;**1**(2):6-36

[2] Fu W, Ma J, Chen P, Chen F. Remote sensing satellites for digital earth. In: Manual of Digital Earth. Singapore: Springer; 2020. pp. 55-123

[3] Christopherson JB, Chandra SNR, Quanbeck JQ. 2019 Joint Agency Commercial Imagery Evaluation—Land Remote Sensing Satellite. Reston, VA: U.S. Geological Survey; 2019

[4] Zhang Y, Kerle N. Satellite remote sensing for near-real-time data collection. Geospatial Information Technology for Emergency Response. 2008;**6**:75-102

[5] Sajjad H, Kumar P. Future challenges and perspective of remote sensing technology. In: Applications and Challenges of Geospatial Technology. Cham: Springer; 2019. pp. 275-277

[6] Chi M, Plaza A, Benediktsson JA, Sun Z, Shen J, Zhu Y. Big data for remote sensing: Challenges and opportunities. Proceedings of the IEEE. 2016;**104**(11): 2207-2219

[7] Zhu L, Suomalainen J, Liu J, Hyyppä J, Kaartinen H, Haggrén H. A Review: Remote Sensing Sensors. London, United Kingdom: IntechOpen; 2018

[8] Rast M, Painter TH. Land observation imaging spectroscopy for terrestrial systems: An overview of its history, techniques, and applications of its missions. Surveys in Geophysics. 2019; **40**(3):303-331

[9] Earth Observations for Official Statistics Satellite Imagery and Geospatial Data Task Team Report. 2017. Available from: https://unstats.un.org/bigdata/taskteams/satellite/UNGWG_Satellite_Task_Team_Report_WhiteCover.pdf

[10] Nagne AD, Dhumal RK, Vibhute AD, Nalawade DB, Kale KV, Mehrotra SC. Advances in land use classification of urban areas from hyperspectral data. Management. 2018; **12**:21

[11] Saah D, Tenneson K, Matin M, Uddin K, Cutter P, Poortinga A, et al. Land cover mapping in data scarce environments: Challenges and opportunities. Frontiers in environmental. Science. 2019;**7**:150

[12] Hurskainen P, Adhikari H, Siljander M, Pellikka PKE, Hemp A. Auxiliary datasets improve accuracy of object-based land use/land cover classification in heterogeneous savanna landscapes. Remote Sensing of Environment. 2019;**233**:111354

[13] Zhong Y, Ma A, Soon Ong Y, Zhu Z, Zhang L. Computational intelligence in optical remote sensing image processing. Applied Soft Computing. 2018;**64**:75-93

[14] Alshari EA, Gawali BW. Development of classification system for LULC using remote sensing and GIS. Global Transitions Proceedings. 2021; **2**(1):8-17

[15] Singh RK, Sinha VSP, Joshi PK, Kumar M. A multinomial logistic model-based land use and land cover classification for the south Asian Association for Regional Cooperation nations using moderate resolution imaging Spectroradiometer product.

Environment, Development, and Sustainability. 2021;**23**(4):6106-6127

[16] Paul S, Saxena KG, Nagendra H, Lele N. Tracing land use and land cover change in peri-urban Delhi, India, over 1973–2017. Environmental Monitoring and Assessment. 2021; **193**(2):1-12

[17] Alshari EA, Gawali BW. Evaluation of the potentials and challenges of land observation satellites. Global Transitions Proceedings. Elsevier B.V. ScienceDirect. 2021

[18] Khwarahm NR. Spatial modeling of land use and land cover change in Sulaimani, Iraq, using multitemporal satellite data. Environmental Monitoring and Assessment. 2021;**193**(3):1-18

[19] Makwinja R, Kaunda E, Mengistou S, Alamirew T. Impact of land use/land cover dynamics on ecosystem service value—A case from Lake Malombe. Southern Malawi. Environmental Monitoring and Assessment. 2021; **193**(8):1-23

[20] Nayak S. Land use and land cover change and their impact on temperature over Central India. Letters in Spatial and Resource Sciences. 2021; **2021**:1-12

[21] Sarif MO, Gupta RD. Spatiotemporal mapping of land use/land cover dynamics using remote sensing and GIS approach: A case study of Prayagraj City, India (1988–2018). Environment, Development, and Sustainability. 2021; **2021**:1-33

[22] Xie FD, Wu X, Liu LS, Zhang YL, Paudel B. Land use and land cover have change within the Koshi River Basin of the Central Himalayas since 1990. Journal of Mountain Science. 2021;**18**(1): 159-177

[23] Sang X, Guo Q, Wu X, Xie T, He C, Zang J, et al. The effect of DEM on the land use/cover classification accuracy of Landsat OLI images. Journal of the Indian Society of Remote Sensing. 2021; **2021**:1-12

[24] Bhattacharya RK, Das Chatterjee N, Das K. Land use and land cover change and its resultant erosion susceptible level: An appraisal using RUSLE and logistic regression in a tropical plateau basin of West Bengal, India. Environment, Development and Sustainability. 2021;**23**(2):1411-1446

[25] Angessa AT, Lemma B, Yeshitela K. Land-use and land-cover dynamics and their drivers in the central highlands of Ethiopia with special reference to the Lake Wanchi watershed. GeoJournal. 2021;**86**(3):1225-1243

[26] Navin MS, Agilandeeswari L. Multispectral and hyperspectral images based land use/land cover change prediction analysis: An extensive review. Multimedia Tools and Applications. 2020;**79**(39):29751-29774

[27] Dibs H, Hasab HA, Al-Rifaie JK, Al-Ansari N. An optimal approach for land-use/land-cover mapping by integration and fusion of multispectral landsat OLI images: Case study in Baghdad, Iraq. Water, Air, & Soil Pollution. 2020; **231**(9):1-15

[28] Kaya İA, Görgün EK. Land use and land cover change monitoring in Bandırma (Turkey) using remote sensing and geographic information systems. Environmental Monitoring and Assessment. 2020;**192**(7):1-18

[29] Xu X, Shrestha S, Gilani H, Gumma MK, Siddiqui BN, Jain AK. Dynamics and drivers of land use and land cover changes in Bangladesh.

Regional Environmental Change. 2020; **20**(2):1-11

[30] MohanRajan SN, Loganathan A, Manoharan P. Survey on Land Use/Land Cover (LU/LC) change analysis in remote sensing and GIS environment: Techniques and challenges. Environmental Science and Pollution Research. 2020;**27**:29900-29926

[31] Rojas F, Rubio C, Rizzo M, Bernabeu M, Akil N, Martín F. Land use and land cover in irrigated drylands: A long-term analysis of changes in the Mendoza and Tunuyán River basins, Argentina (1986–2018). Applied Spatial Analysis and Policy. 2020;**13**(4):875-899

[32] Saddique N, Mahmood T, Bernhofer C. Quantifying the impacts of land use/land cover change on the water balance in the afforested River Basin. Pakistan. Environmental Earth Sciences. 2020;**79**(19):1-13

[33] Ekumah B, Armah FA, Afrifa EK, Aheto DW, Odoi JO, Afitiri AR. Assessing land use and land cover change in coastal urban wetlands of international importance in Ghana using intensity analysis. Wetlands Ecology and Management. 2020;**28**(2):271-284

[34] Rindfuss RR, Walsh SJ, Turner BL, Fox J, Mishra V. Developing a science of land change: Challenges and methodological issues. Proceedings of the National Academy of Sciences. 2004; **101**(39):13976-13981

[35] Hu Y, Li W, Wright D, Aydin O, Wilson D, Maher O, Raad M. Artificial intelligence approaches. 2019. arXiv preprint arXiv:1908.10345

[36] Yuan H, Van Der Wiele CF, Khorram S. An automated artificial neural network system for land use/land cover classification from Landsat TM imagery. Remote Sensing. 2009;**1**(3): 243-265

[37] Girma R, Fürst C, Moges A. Land use land cover change modeling by integrating artificial-neural-network with cellular Automata-Markov chain model in Gidabo river basin, main Ethiopian rift. Environmental Challenges. Elsevier B.V. ScienceDirect. 2021:100419

[38] Alqadhi S, Mallick J, Balha A, Bindajam A, Singh CK, Hoa PV. Spatial and decadal prediction of land use/land cover using multi-layer perceptron-neural network (MLP-NN) algorithm for a semi-arid region of Asir, Saudi Arabia. Earth Science Informatics. 2021;**14**(3): 1547-1562

[39] Ramdani F, Setiawan B, Rusydi A, Furqon M. An Artificial Neural Network Approach to Predict the Future Land Use Land Cover of Great Malang Region. Indonesia; 2021

[40] Sahithi VS, Nehru J, Krishna IVM, Movement SV, Giridhar MVS. Hyperspectral Data Classification Algorithms for delineation of LULC classes. 2021

[41] Talukdar S, Eibek KU, Akhter S, Ziaul S, Islam ARMT, Mallick J. Modeling fragmentation probability of land-use and land-cover using the bagging, random forest and random subspace in the Teesta River Basin, Bangladesh. Ecological Indicators. 2021;**126**:107612

[42] Verma D, Jana A. LULC classification methodology based on simple Convolutional Neural Network to map complex urban forms at finer scale: Evidence from Mumbai. 2019. arXiv preprint arXiv:1909.09774

[43] Alshari EA, Gawali BW. Analysis of machine learning techniques for

sentinel-2A satellite images. Journal of Electrical and Computer Engineering. 2022;**2022**:16

[44] Alshari EA, Gawali BW. Modeling for land use changes of Sana'a City of Yemen using MOLUSCE. Journal of Sensors. 2022;**2022**:15. DOI: 10.1155/2022/7419031

[45] Alshari EA, Gawali BW. Classification of land use land cover using artificial intelligent (ANN-RF). Journal: Frontiers in Artificial Intelligence, Section Machine Learning and Artificial Intelligence. 2022; **2022**:15. Available from: https://10.3389/frai.2022.964279

[46] Li Z, Guo J, Ji B, Wan X, Zhang S. A review of marine gravity field recovery from satellite altimetry. Remote Sensing. 2022;**14**(19):4790

[47] Quilfen Y, Piolle JF, Chapron B. Towards improved analysis of short mesoscale sea level signals from satellite altimetry. Earth System Science Data. 2022;**14**(4):1493-1512

[48] Available from: https://ggos.org/item/satellite-altimetry/

[49] Yang L, Lin L, Fan L, Liu N, Huang L, Xu Y, et al. Satellite altimetry: Achievements and future trends by a Scientometrics analysis. Remote Sensing. 2022;**14**(14):3332